D0082783

Solar Plasma Geomagnetism and Aurora

P-21
P
F.C.

SYDNEY CHAPMAN
Geophysical Institute, University of Alaska
and High Altitude Observatory
Boulder, Colorado

GORDON AND BREACH
Science Publishers

New York · London

EDITOR'S PREFACE

Seventy years ago when the fraternity of physicists was smaller than the audience at a weekly physics colloquium in a major university, a J. Willard Gibbs could, after ten years of thought, summarize his ideas on a subject in a few monumental papers or in a classic treatise. His competition did not intimidate him into a muddled correspondence with his favorite editor nor did it occur to his colleagues that their own progress was retarded by his leisurely publication schedule.

Today the dramatic phase of a new branch of physics spans less than a decade and subsides before the definitive treatise is published. Moreover, modern physics is an extremely interconnected discipline and the busy practitioner of one of its branches must be kept aware of breakthroughs in other areas. An expository literature which is clear and timely is needed to relieve him of the burden of wading through tentative and hastily written papers scattered in many journals.

To this end we have undertaken the editing of a new series, entitled *Documents on Modern Physics,* which will make available selected reviews, lecture notes, conference proceedings, and important collections of papers in branches of physics of special current interest. Complete coverage of a field will not be a primary aim. Rather, we will emphasize readability, speed of publication, and importance to students and research workers. The books will appear in low-cost paper-covered editions, as well as in cloth covers. The scope will be broad, the style informal.

From time to time, older branches of physics come alive again, and forgotten writings acquire relevance to recent developments. We expect to make a number of such works available by including them in this series along with new works.

ELLIOTT W. MONTROLL
GEORGE H. VINEYARD

150 Fifth Avenue, New York, New York 10011

Library of Congress Card Catalog Card Number 64-21214

PRINTED IN THE UNITED STATES OF AMERICA

CONTENTS

Preface

This little book is based on lectures given in July 1962 at the twelfth annual session of the Les Houches Summer School of Theoretical Physics, sponsored by the University of Grenoble and directed by Mme. C. DeWitt; with the work of the other lecturers at the School it was first published in GEOPHYSICS: The Earth's Environment, under the chapter title Solar Plasma, Geomagnetism and Aurora. As the aurora is only briefly treated, reference to it is omitted from the title of this book.

It may be regarded as a supplement to *Geomagnetism* (Chapman and Bartels, 1940). But unlike that book, it does not attempt to give a general survey of the topics considered. As is permissible in a short course of lectures, one of a group of courses, the treatment is selective. The evolution of knowledge and theory concerning magnetic storms is described with special reference to the work of my colleagues and myself, and some later developments therefrom. For accounts of alternative theories and later developments, particularly as regards balloon, rocket and satellite exploration of the magnetosphere and inter-planetary space, and of the flux of particles and x-rays into the atmosphere and beyond, reference should be made to the books and other writings by Akasofu, Alfvén, Anderson, Arnoldy, Axford, Bartels, Beard, Bhavsar, Cahill, Cain, Chamberlain, Cole, Davis, Dessler, Dungey, Fejer, Ferraro, Freeman, Freier, Galperin, Gold, Heppner, Hines, Hoffman, Hulburt, Hultquist, Jacobs, Kato, Kellogg, Kern, Krassovskii, Maehlum, Martyn, McIlwain, Nagata, Neugebauer, Obayashi, O'Brien, Omholt, Parker, Paton, Piddington, Reid, Smith, Snyder, Sonett, Spreiter, Stoffregen, Sugiura, Van Allen, Vernov, Vestine, Webber, Winckler and others.

1. Geomagnetism and Related Phenomena

1.1. Historical

Robert Norman, a London ship's instrument maker, published a small pamphlet in 1576 describing his important discovery that a needle, perfectly balanced while non-magnetic, did not remain horizontal after it was magnetized. Mounting the needle so that it was free to turn in the vertical plane of magnetic north, he found that the north end dipped by about 70°.

At that time William Gilbert, physician to Queen Elizabeth, and a contemporary of Shakespeare, was spending much of his leisure in experiments on magnetism and static electricity. In 1600 he published, in Latin, the great treatise *de Magnete*, in which he reviewed the earlier writings on these subjects, refuted much pseudo-science, and described his own discoveries. By tiny magnets he explored the surface field of a spherical lodestone (magnetite); he traced on it the lines of the tangential component of magnetic force, as had been done more than three centuries earlier by Petrus Peregrinus (1269), who saw that they converged to two opposite points, which he called magnetic poles. Gilbert also noted how the tiny magnets dipped at different angles in different "latitudes" relative to these poles. Remembering Norman's discovery, his imagination bridged the great difference of scale. Initiating the science of *geo*magnetism, he wrote:

Magnus magnes ipse est globus terrestris

(The earth globe itself is a great magnet)

The oldest known consequence of the earth's magnetism is its directive influence on the compass—still of great importance for navigation by sea and in the air. This was known and used for centuries before Gilbert saw that the cause lies within the earth and not, as many had supposed, in the heavens.

In 1635 Gellibrand showed that the field slowly changes. At London the compass moved steadily westward for 220 years, from 11° E in 1580 to 24° W in 1800; the dip also changed, increasing to $74\frac{1}{2}$° by 1700; since then it has gradually decreased to its present value, 66°, which may prove to be a minimum. During the last century the earth's magnetic moment decreased by 5 per cent; now it may be increasing again.

In the years 1698 to 1700 Edmund Halley made the first ocean magnetic survey, in the north and south Atlantic Oceans. In 1701 he embodied his observations in the first oceanic magnetic chart. A year later, having collected many observations of the compass direction made by other mariners, he published the first *world* magnetic chart. Later studies of ships' logs have enabled magnetic charts to be drawn, for parts of the globe, for as far back as 1550. Such charts are now published by the chief hydrographic offices, and revised every five years. A new world magnetic survey, more complete than any ever before made, will be undertaken internationally as a deferred part of the International Geophysical Year (IGY), during the years 1964–65. This and other geophysical programs to be undertaken during that period constitute an enterprise called the International Quiet Sun Year, (IQSY). Magnetic observations will be made on land, at sea, and by aircraft, satellites, and rockets. They will lead to new world magnetic charts for the epoch 1965.0. Owing to the slow secular variation of the earth's magnetism, the production of such charts is a never-ending task.

Going backward in time, again, by more than two centuries, the next important step in the science of the earth's magnetism was the discovery that it continually undergoes *transient* changes. This was announced in 1722 by Graham, an eminent London clock-maker. It was based on his close and long continued observations of the small movements of a compass needle. These included regular daily variations, on which at times additional larger and irregular changes are superposed. Such irregular changes are called *magnetic disturbance*. Thus Graham first made the distinction between geomagnetic quiet or calm, and magnetic disturbance or activity. When a magnetic disturbance is considerable, it is called a *magnetic storm*—a name introduced by Humboldt.

In 1740 Celsius, of Upsala, Sweden, began similar studies; comparing notes by letter, he and Graham found in 1741 that magnetic disturbance was often simultaneous at the two places. This showed that magnetic "weather" is less local than ordinary weather. Magnetic storms, indeed, are worldwide. Celsius also found, in 1741, that auroras are correlated with magnetic disturbance. In 1770 his countryman Wilcke at Stockholm found that the auroral rays lie along the earth's magnetic field lines.

General magnetic *theory* was greatly advanced in 1824 by the French mathematician Poisson. He defined the concept of the magnetic dipole and of intensity of magnetization, and extended the theory of the potential, already developed for gravity and electricity, to magnetism.

In 1832 the great German scientist Gauss showed how to measure magnetic intensity in absolute units. In 1834 he set up at Göttingen the first magnetic observatory at which all three magnetic elements could be measured—by eye readings, with mirror and scale; photographic recording did not come until 1847, when it was installed at Greenwich by Airy. Similar magnetic observatories were soon set up at other places, and it began to be possible to obtain a synoptic picture of the world distribution of the transient magnetic changes. During the IGY there were over 150 magnetic observatories, widely distributed over the globe.

In 1826 a young German apothecary of Dessau, Samuel Heinrich Schwabe, at the age of thirty-six, was stimulated by a friend, who knew of his astronomical interests, to take up the neglected regular observation of sunspots. Schwabe was a most regular and consistent observer; later it was said of him that "the sun never rose unclouded above the horizon of Dessau without being encountered by Schwabe's imperturbable telescope". After twelve years' observations he published his spot counts in the German *Astronomische Nachrichten* (News). He already saw that they suggested a variation with a period of about ten years, but he made this comment only in 1843, when in the same journal he published a further five years' observations. He gave both annual spot counts and annual numbers of spotless days; both showed in inverse ways the cycle of rising and falling activity. Figure 1 shows the sunspot numbers from 1700 A.D. until 1960. Schwabe continued his observations, and in 1857 when he was awarded the Gold Medal of the Royal Astronomical Society the president remarked: "Twelve years he spent to satisfy himself, six more years to satisfy, and still thirteen more to convince, mankind" of his discovery; and that thus "the energy of one man has revealed a phenomenon that had eluded the suspicion of astronomers for 200 years". But for some years after 1843 this discovery of the *solar* or *sunspot* cycle remained little noticed.

In 1851 Schwabe's recognition of the existence of the sunspot cycle was publicized by Humboldt in his famous book *Cosmos*, and within a year Sabine and others announced that the same cycle was shown in the transient magnetic variations (Fig. 2). Thus it became clear that whereas ordinary weather is influenced

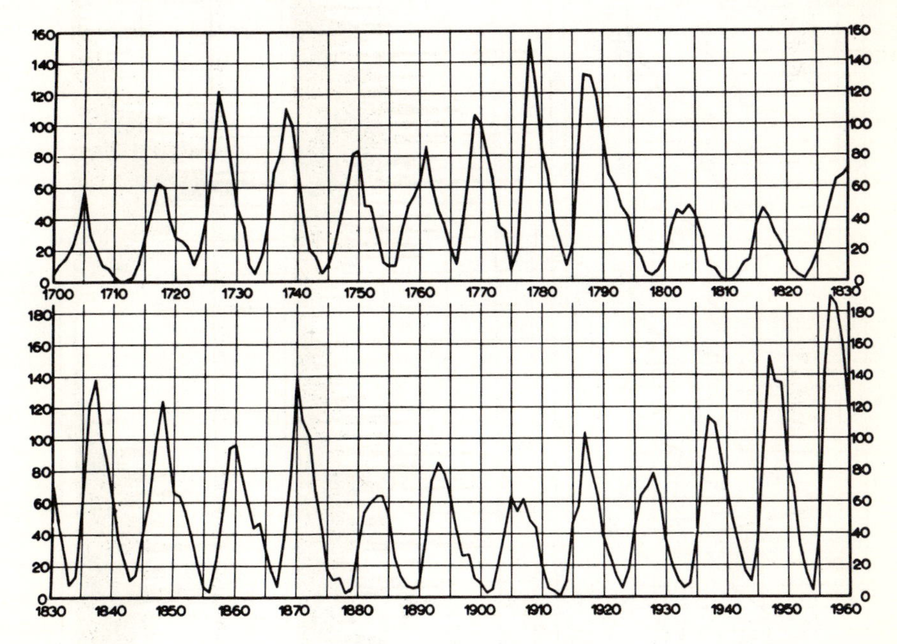

Fig. 1.1. Sunspot numbers, annual means, 1700–1960 (after Waldmeier). (Waldmeier 1961)

by the changing *geometrical* relationships of the earth and sun, the *magnetic* "weather" is influenced by intrinsic changes on the sun's surface.

In 1860 Elias Loomis, professor of natural philosophy at Yale, drew the first diagram of the auroral zone, and pointed out that its oval form, not centered on the geographical pole, somewhat resembles that of the lines of equal magnetic dip—yet another connection between the aurora and the earth's magnetic field. In 1873–4 Hermann Fritz, then professor of physics at Zürich, published a great auroral catalog, and used it to give a map showing lines—which he called isochasms—of equal frequency of auroral visibility.

By 1880 it had become clear that the small regular daily magnetic changes are greater by 50 per cent or more at sunspot maximum than at sunspot minimum.

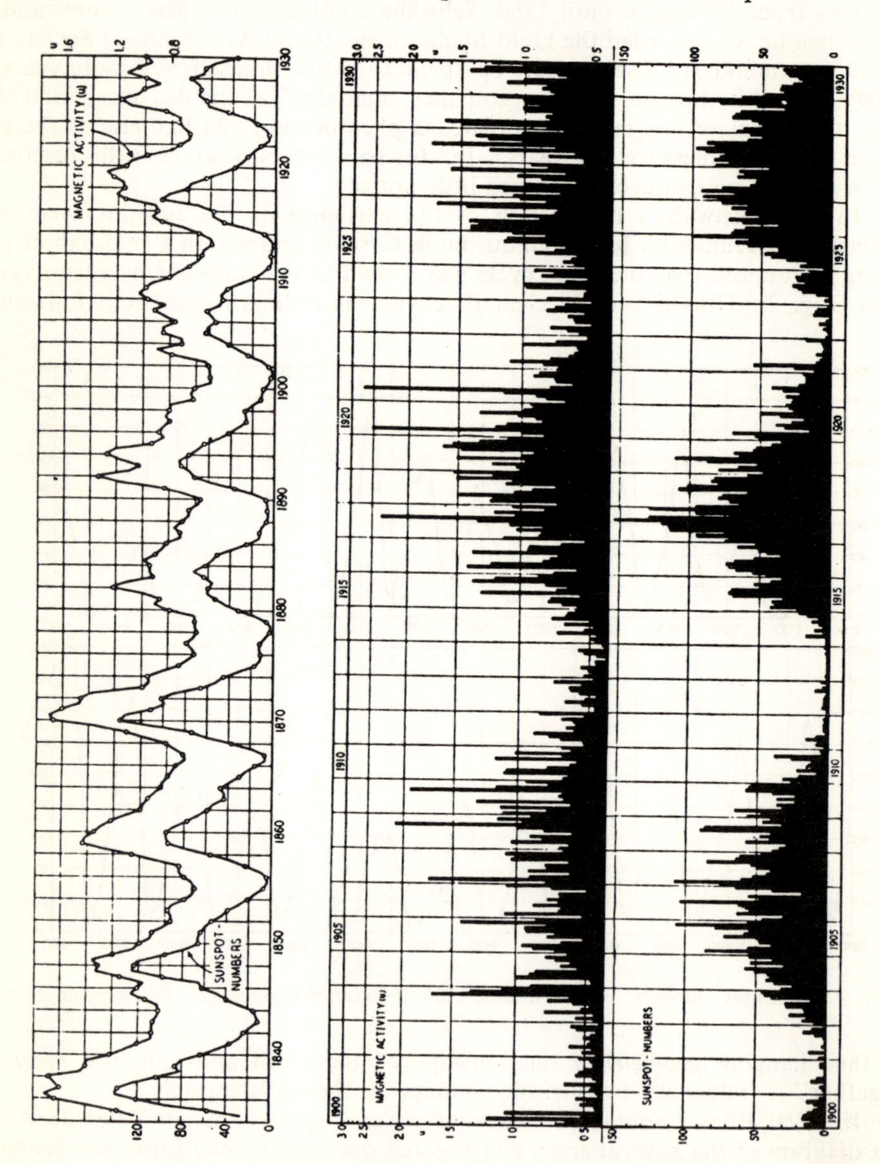

Fig. 1.2. Sunspot numbers and geomagnetic disturbance (as measured by u, the inter-diurnal variability); annual means (above), monthly means (below). (Bartels 1932, p.13; also Geomagnetism, p.369.)

To the Scottish physicist Balfour Stewart this was a clear pointer to the source of these variations. In his article on Terrestrial Magnetism in the 9th (1882) edition of the *Encyclopedia Britannica*, he showed by sound arguments that their cause must be electric currents flowing in the upper atmosphere, and that the air there must be rendered electrically conducting by solar action. This was the first recognition of what is now called the ionosphere. Stewart concluded that the electric currents must be induced by dynamo action of airflow across the geomagnetic field. Their own varying field induces secondary currents within the earth. The field of these earth currents appreciably modifies the daily magnetic variations observed at the earth's surface.

These studies showed that the upper atmosphere must be more conducting, that is, more ionized, by day than by night, in summer than in winter, and at sunspot maximum than at sunspot minimum.

Twenty years later (1902) the existence of the ionosphere was recognized anew, by two men independently—Heaviside and Kennelly—from quite different evidence. From Marconi's success in 1901 in sending radio waves from Cornwall to Newfoundland, they concluded that the waves must be made to follow the curve of the earth by an electrically conducting layer in the upper atmosphere. After another twenty years radio waves began to be a valuable tool for the exploration of the ionosphere from the ground. Appleton and others in England, and Breit and Tuve and others in the U.S.A., were the pioneers in such researches, which soon revealed the layered structure of the ionosphere—E, F1, F2.

1.2. The external field of a uniformly magnetized sphere

Gilbert experimented with a sphere of magnetite, presumably uniformly magnetized: that is, each element dv of its volume had a dipole moment $\mathbf{J}dv$, where $\mathbf{J}$, the intensity of magnetization, was the same throughout*. The external field of such a sphere, of radius a, is the same as that of a point dipole of moment $\mathbf{M}$ at O, the center of the sphere; and

$$\mathbf{M} = \frac{4}{3}\pi a^3\mathbf{J} \tag{1.1}$$

Of course $\mathbf{J}$ and $\mathbf{M}$ are vectors, in the same direction. The diameter BA in that direction is called the *magnetic axis* of the sphere, and B and A are called the *magnetic poles*, A being the positive and B the negative pole. By convention, a positive pole is the one that, in a compass, turns northward over most of the earth. The plane through O perpendicular to BA is called the *magnetic equatorial plane*, and the circle in which it cuts the sphere is called the *magnetic equator*. *Magnetic latitude* Φ and co-latitude or north polar (angular) distance $\Theta(= 90° - \Phi)$ are reckoned relative to this equator. The semicircles joining B and A are called the *magnetic meridians*; one is chosen as the zero of *magnetic longitude* Λ.

The magnetic potential V of the dipole, at any point P whose position vector relative to O is $\mathbf{r}$, is given by

$$V = (\mathbf{M} \cdot \mathbf{r})/r^3 = -(M\cos\Theta)/r^2, \tag{1.2}$$

*Symbols in heavy type denote vectors; the same symbols in lighter type denote the positive magnitude of the vector.

where Θ (the colatitude) denotes the angle BOP (Fig. 3); V is also the potential of the *external* field of the uniformly magnetized sphere. The intensity $\mathbf{F}$ of the field is given by

$$\mathbf{F} = -\text{ grad } V. \tag{1.3}$$

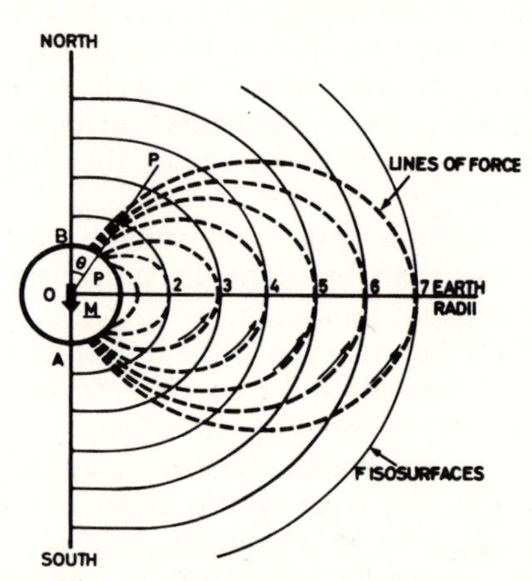

Fig. 1.3. Lines of force (– – –) and lines of equal field intensity, for a dipole field or the external field of a uniformly magnetized sphere.

The radial component Z of $\mathbf{F}$, reckoned positive when inward, is given by

$$Z = \partial V/\partial r = (2M \cos \Theta)/r^3. \tag{1.4}$$

The vector component $\mathbf{H}$ tangential to the sphere (r = constant) through P is directed along the meridian through P, towards B. Its strength H and the magnetic dip or inclination I are given by

$$H = \partial V/r\partial\Theta = (M \sin \Theta)/r^3, \tag{1.5}$$

$$\tan I = Z/H = 2 \cot \Theta. \tag{1.6}$$

Because this is independent of r, the dip is the same at all points along any radius OP. For the earth (6) is used to define (p.18) a *dip* latitude. Also

$$F = (H^2 + Z^2)^{1/2} \tag{1.7}$$

At the pole B, $F = Z$, and at the equator, $F = H$. Let H_0 denote the equatorial value of H at the surface of the sphere ($r = a$); the *polar* value of F at the surface is $2H_0$. In terms of H_0,

$$H = H_0(a/r)^3 \sin \Theta, \quad Z = 2H_0(a/r)^3 \cos \Theta,$$

$$F = H_0(a/r)^3(1 + 3 \cos^2\Theta)^{1/2}. \tag{1.8}$$

In terms of the intensity of magnetization $\mathbf{J}$ of the sphere,

$$H_0 = M/a^3 = \frac{4}{3}\pi J. \tag{1.9}$$

The lines of force of the dipole **M**, and of the external field of the sphere, are given by $dr/Z = rd\Theta/H$; hence their equation is

$$r = ka \sin^2\Theta. \tag{1.10}$$

Here k is a parameter of the particular line; clearly ka is the distance at which the line crosses the equatorial plane $\Theta = 90°$. The points where it meets the sphere are given by

$$\Theta = \Theta_0, \quad \Theta = \pi - \Theta_0, \quad \text{where} \quad \sin\Theta_0 = 1/k^{1/2}$$

The lines of force, together with the isosurfaces of F (on which F is constant), given by

$$r = k'a(1 + 3\cos^2\Theta)^{1/6}, \tag{1.11}$$

indicate the direction and strength of the field at each point (Fig. 3). The parameter k' of the F isosurface is thus related to the value of F for the surface:

$$k' = (H_0/F)^{1/3}.$$

Table I gives for various lines of force, that cross the equator at distance ka, the values of Θ_0, and also the latitudes $\pm\Phi_0$, of the two *ends* of the line, on the sphere; these two points are said to be magnetically "conjugate". The Table also gives the value of the dip, $+I$, at these points. As Gilbert found, the dip increases from zero at the equator to 90° at B.

TABLE 1.1

Lines of Force

k:	1	2	3	4	5	6	7
Θ_0:	90°	45°	35.3°	30°	26.6°	24.1°	22.2°
$\pm\Phi_0$:	0	45	54.7	60	63.4	65.9	67.8
$\pm I$:	0	63.4	70.5	73.9	76.0	77.4	78.5
k:	8	9	10	11	12	13	14
Θ_0:	20.7°	19.5°	18.4°	17.5°	16.8°	16.1°	15.5°
$\pm\Phi_0$:	69.3	70.5	71.6	72.5	73.2	73.9	74.5
$\pm I$:	79.3	80.0	80.5	81.0	81.4	81.8	82.1

Table 2 gives the values of F/H_0 for the F isosurfaces that cross the equatorial plane at similar distances $k'a$ from O.

TABLE 1.2

F Isosurfaces

k':	1	2	3	4	5
F/H_0:	1	0.125	0.0370	0.0156	0.008
k':	6	7	8	9	10
F/H_0:	0.00463	0.00292	0.00195	0.00137	0.00100

The angle made with the outward radial direction by the line of force whose parameter is k, at the point r, Θ, is $\pi - \varphi$, where

$$\tan\varphi = H/Z = \tfrac{1}{2}\tan\Theta.$$

Thus the line of force there makes the angle χ given by $(\pi - \varphi) - \Theta$ with the fixed direction of the dipole axis. The radius of curvature ρ of the line at r, Θ is $ds/d\chi$; by (10),

$$(ds/d\Theta)^2 = (dr/d\Theta)^2 + r^2 = (kaC\sin\Theta)^2,$$

where

$$C^2 = 1 + 3\cos^2\Theta. \qquad (1.12)$$

Also

$$|d\chi/d\Theta| = 1 + d\varphi/d\Theta = 3(1 + \cos^2\Theta)/C^2.$$

Hence

$$\rho = ds/d\chi = (ds/d\Theta)/(d\chi/d\Theta) = (kaC^3\sin\Theta)/3(1 + \cos^2\Theta) \qquad (1.13)$$

1.3. The magnetic field at the earth's surface

The intensity **F** of the earth's magnetic field at any point P can be specified by its downward vertical component Z and its vector horizontal component **H**, or by **H** and the angle I by which **F** dips below the horizontal; and

$$F^2 = H^2 + Z^2, \quad \tan I = Z/H; \qquad (1.14)$$

cf. (6,7). The positive magnitudes of **H** and Z are often denoted by H. F. and V. F., or H and V, and called the horizontal and vertical "force".

On the earth, the direction of **H** is specified by the angle D between **H** and geographical north; D is called the magnetic (or compass) declination and is reckoned positive if eastward. The northward and eastward components of **H** are denoted by X, Y, respectively, and

$$\tan D = Y/X, \quad X = H\cos D, \quad Y = H\sin D, \quad H^2 = X^2 + Y^2. \qquad (1.15)$$

The seven quantities F, Z, H, I, D, X, Y are called *magnetic elements,* and any set of three independent elements serves to specify **F**, such as

$$F, I, D; \quad H, I, D; \quad H, Z, D; \quad X, Y, Z. \qquad (1.16)$$

The elements F, H, Z, I are called *intrinsic,* because their only reference to direction is to the natural direction characteristic of P, namely the vertical. The other three elements, D, X, Y, are called relative, because they are defined *relative* to the geographical (or rotation) axis NS, whose northern and southern (geographical) poles N, S have no necessary relation to the geomagnetic field.

In the science of geomagnetism, which owes so much to Gauss, magnetic intensity is expressed in gauss units (Γ), despite internationally agreed use by physicists and engineers of the name oersted as the unit of intensity and gauss as the unit of magnetic induction. For measurements of the earth's field in air, the difference is insignificant. A smaller unit, the gamma (γ), is also used, especially in connection with the secular and transient magnetic variations:

$$1\gamma = 10^{-5}\Gamma. \tag{1.17}$$

1.3a. Isomagnetic charts

The field distribution over the earth's surface is determined by magnetic surveys. They usually extend over months or years, during which measurements of one or more elements are made at many points. The observations, made on different dates, are later corrected to a common epoch, using the best available value of the secular variation for each element. This involves some uncertainty.

The survey results are plotted on isomagnetic charts, namely maps on which isomagnetic lines (or isolines) are drawn. Usually one chart relates to one element only, though sometimes the vector lines of **H**, called magnetic meridians (cf. §2), are also drawn on an isomagnetic chart, particularly that for the dip or inclination I. These **H** lines give the clearest indication of the general direction of **H**, but for practical use, e.g., by navigators, the D isomagnetic chart is more convenient. The D chart and its lines are called *isogonic* (the lines are also called *isogones*); the I chart and its lines are called *isoclinic* (the lines are also called *isoclines*). The F chart is called *isodynamic*. Along each isoline, the element concerned has a constant value. The chart relates to a particular epoch. The first isomagnetic charts were the isogonic charts for 1700 and 1701 by Halley.

For the field of a uniformly magnetized sphere the isolines for the intrinsic elements F, Z, H, I are simple—they are the circles of constant magnetic latitude (§2). If the magnetic axis is taken as that of reference for the direction of **H**, then $X = H$, $Y = 0$, $D = 0$. But if some other axis NS is used, the isogonic chart has four singular points, A, B, N, S; this is because D takes all values round any small closed curve that encircles any of these points; thus all isogones, from $D = 0$ to $D = 360°$, meet at each of these four points.

The earth's external field somewhat resembles that of a point dipole or uniformly magnetized sphere. But this is only a first approximation. There are regional departures from a dipole field. The isoclinic lines are simple closed curves that shrink to points P_N, P_S in the northern and southern hemispheres respectively. But these curves are not circles in parallel planes, as in §2; and their limiting points P_N, P_S are not at opposite ends of a diameter, like B and A in Fig. 3. These points, where the needle dips vertically and the compass takes no direction, are called the (earth's) *magnetic poles* or *dip poles*. Their location is not accurately known. Indeed their positions shift daily and seasonally by small amounts, because of the transient variations. Approximate positions for 1945 were:

P_N 76.0° N; 102.0° W

P_S 68.2 S; 145.4 E

In 1904 the estimated position of P_N was 70.0° N, 96.7° W, or 600 km from the present position.

The two dip poles are not antipodal; the antipode of P_S is about 2000 km from P_N. The line $P_N P_S$ misses the earth's center by about 1000 km.

1.3b. The magnetic potential and the dipole field

From the H and D or X and Y isomagnetic charts it is possible to calculate the earth's magnetic potential V at any point P on the earth's surface; thus

$$V = \int_{P_0}^{P} \mathbf{H} \cdot \mathbf{ds}, \tag{1.18}$$

where the integration is to be made along any path on the surface, from an arbitrarily chosen point P_0 to P.

By a mathematical process called spherical harmonic analysis, V can be expressed as a sum of terms of the type

$$P_n^m(\cos\theta)(a_n^m \cos m\lambda + b_n^m \sin m\lambda), \tag{1.19}$$

where $90°-\theta$ is the N latitude φ, and λ is the east longitude, of P.

Gauss was the first to apply this process to geomagnetic data; he used the charts for D(1833), I(1836) and F(1837). Thus he was able to confirm and extend Gilbert's conclusion that the earth's field originates from within. Any small part that has its source above the earth is at most of order 0.1 per cent of the whole, and cannot as yet be calculated by spherical harmonic analysis from the chart data.

The three main terms in V at the surface are proportionate to $\cos\theta$, $\sin\theta\cos\lambda$, and $\sin\theta\sin\lambda$. The $\cos\theta$ term is the largest; it represents the field of a dipole at O, the earth's center, in the direction NS. The other two terms represent the fields of dipoles at O, lying in the plane of the geographic equator, respectively, in and perpendicular to the meridian of Greenwich. The three terms together represent the field of a dipole of moment **M** at O, along a diameter BA; this is called the *(magnetic) dipole axis.*

The analysis by Finch and Leaton (1957), for the epoch 1955.0, gave

$$M = 8.07 \times 10^{25}\ \text{gauss cm}^3, \quad H_0 = 0.3120\,\Gamma = 31200\,\gamma. \tag{1.20}$$

Here, as in §1.2, H_0 denotes the surface intensity of the dipole field at the *dipole equator* (in the plane through O normal to BA). The northern and southern ends B,A of the dipole axis are called the *boreal* and *austral* (dipole) poles; their 1955 geographical coordinates were

$$\theta_0 = 11.7°,\ \phi_0 = 78.3°\text{N},\ \lambda_0 = 291.0°\ \text{E (or } 69.0°\text{W)} \tag{1.20a}$$

$$\theta'_0 = 168.3°,\ \phi'_0 = 78.3°\text{S},\ \lambda'_0 = 111.0°\text{E}. \tag{1.20b}$$

A point P with geographic ("*gg*") coordinates θ,ϕ,λ has (magnetic) *dipole* ("dpl") coordinates Θ,Φ,Λ—often called *geomagnetic* ("gm")—given by

$$\cos\Theta = \cos\theta\cos\theta_0 + \sin\theta\sin\theta_0\cos(\lambda - \lambda_0) \tag{1.21}$$

$$\sin\Lambda = \sin\theta\sin(\lambda - \lambda_0)/\sin\Theta. \tag{1.22}$$

The contraction *gm* for these coordinates may conveniently continue in use, or be replaced in time by *dpl*, but the word *geomagnetic* is best reserved for use in its general sense. *IGY Annals*, **8,** 159–214 gives *gg* and *dpl* (or gm) coordinates for the IGY geomagnetic, auroral, ionospheric and cosmic ray stations, taking $\theta_0 = 11.5°$, differing insignificantly from (1.20a).

Tables for conversion from geographic to magnetic dipole coordinates are available (Smithsonian Physical Tables, 9th ed., 1954).

A century ago the moment was about 5 percent greater than now, in almost the same direction.

Finch and Leaton calculated the first 48 harmonic terms in the magnetic potential V for 1955.0. The higher terms, for which, in (19), $n > 1$, represent the departure of the field from that of the dipole **M** at O; they "explain" the deviation of the dip poles P_N,P_S from the dipole poles B,A, and the irregularities of the various isomagnetic lines.

A second approximation to the field is that of a dipole of the same moment **M**, given by (20), situated at an eccentric point O′; it is 340 km from O, in latitude 6.5° N and longitude 161.8° E; the displacement OO′ is nearly perpendicular to BA. The field of this *eccentric dipole* is much stronger at the end of the radius OO′ than at the opposite end of this diameter; the values are 0·371 and 0·270 gauss. Other terms in V contribute regional deviations (of comparable amount) from the field of the *central* dipole M.

1.3c. The origin and secular variation of the field

The best available explanation of the magnetic field of the earth is that it is caused by electric currents flowing in the earth's core, below a depth of 2900 km. Except for a small inner core, the material there is molten, and in a state of slow convection—perhaps caused by non-uniform generation of heat by the decay of irregularly distributed radioactive materials. By dynamo action such flow can induce electromotive forces that can set up and maintain electric currents. We observe the stray field of such currents at the earth's surface; the field in the core may be much stronger, with its lines of force partly wrapped around the axis. Large-scale changing eddies in the convection produce the irregular part of the field, and most of its secular variation.

The magnetic field at the earth's surface has been *measured* only during the last few centuries, at most. But its past history can to some extent be traced by the study of the magnetism of igneous rocks and of sedimentary rocks, on land or in the sea bed. They may retain magnetization corresponding to the strength and direction of the field at the time they solidified or were deposited. The interpretation involves problems of dating and other uncertainties, but there are indications that the earth's dipole field has at times been in a direction nearly opposite to its present direction.

1.4. The magnetic field above the earth's surface

The term 1.19 in the magnetic potential V decreases outwards from the earth's surface as $1/r^{n+1}$. The dipole terms ($n = 1$) decrease as $1/r^2$, and the corresponding force components (§2) as $1/r^3$. The other terms decrease more rapidly; with increasing distance from O, the field approximates ever more closely to that of a dipole at O (or, better, at O′).

Because F decreases at least as rapidly as $1/r^3$, the magnetic field extends much less far (proportionately) into space, than does the gravitational field.

The potential derived from the surface data is valid everywhere above the earth, except in so far as it is modified by electric currents flowing in the ionosphere and beyond. Until we reach the level of the ionosphere, the air is too insulating to permit any appreciable currents to flow. The field lines have been calculated, ignoring the disturbing field of such outside currents, most recently by Vestine and Sibley (1960), using the analysis by Finch and Leaton (1957), who determined the first 48 terms in V; Vestine and Sibley (1960) thus found the locations of numerous pairs of *conjugate points*, the opposite ends (on the surface) of the lines of force.

The course of the field lines above the earth's surface, and the positions of the conjugate points, are nowadays of interest to ionospheric physicists. This is because certain low frequency radio waves, such as the "whistlers" generated by lightning flashes, appear to travel approximately along the lines of force. Their study offers one means of investigating the electrical properties of the space traversed by such waves. The waves go far above the earth—for example, those that start from geomagnetic latitude 60° go out to about 4 earth radii from the earth's center (see Table 1). Hence they can provide information by ground observations about the Van Allen belt regions, otherwise investigated at great cost by satellites.

1.5. Electric currents in and beyond the ionosphere

Days when the transient magnetic variations are "smooth" and regular are called q days, q signifying magnetically *quiet*. The others, d days, are said to be (more or less) magnetically *disturbed* or *active*. The changes indicate the presence of varying fields, superposed on the main field that proceeds from within the earth. The additional fields are those of electric currents that flow in and beyond the ionosphere, and of secondary earth currents induced by the primary varying fields of the external electric currents. The two parts of the transiently changing fields, of primary and induced origin, can be determined by spherical harmonic analysis of the magnetic records made at the many magnetic observatories distributed over the earth. The induced fields reduce the vertical component of the external fields at the earth's surface, and in general they strengthen the horizontal component. About two thirds of the surface variations of the horizontal field are due to the primary external currents, and one third to the induced earth currents.

On q days the magnetic variations proceed mainly according to local solar time; but they also include a part, generally very small, that is controlled partly by the moon. The two parts are called the *solar daily* and *lunar daily* magnetic variations. They, and the fields of which they are the manifestation, are denoted by Sq and L (S for solar and L for lunar, and q for quiet). Both are caused by electric currents flowing in the ionosphere, mainly in the E layer. As was first inferred by Stewart, they are the result of dynamo action of airflow across the geomagnetic field. From the magnetic records it is possible to determine the pattern and approximate intensity of the currents, but not the height at which they flow. This, however, can be inferred from a study of the distribution of the number density n_e of electrons and n_i of ions in the ionosphere, which enables the electric conductivity to be determined, as a function of height and position on the earth with respect to the sun.

1.5a. The Sq electric current system

Figure 1.4 illustrates the *Sq* ionospheric currents in a sunspot minimum year, at the equinoxes and at the June solstice. The current-lines show the direction of flow, over the day hemisphere (as viewed from the sun) on the left, and over the night hemisphere on the right. The lines are spaced so that 10,000 amperes flow between adjacent lines. At the equinoxes about 65,000 amperes flow in each of two daytime circuits, north and south of the equator—anticlockwise and clockwise, respectively. In June the northern circuit is enhanced and enlarged, spreading partly into the southern hemisphere, where the other circuit is much diminished. At sunspot maximum the currents have a similar pattern, but are more intense,

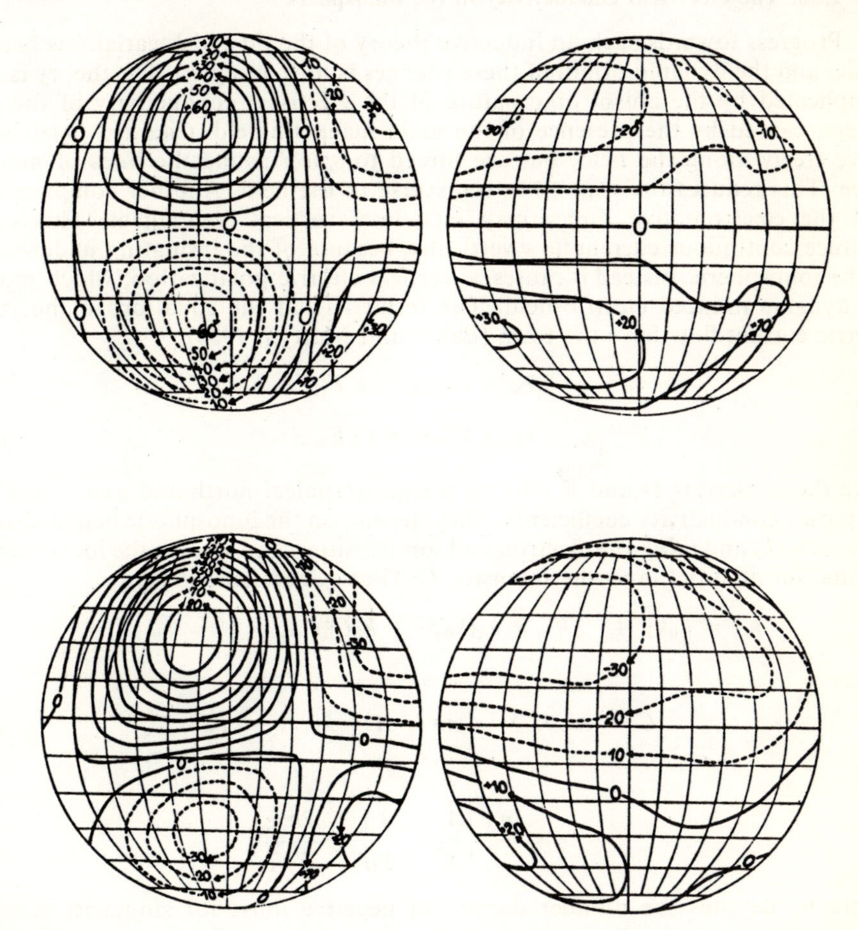

Fig. 1.4. The overhead electric current-systems corresponding to *Sq* (the solar daily magnetic variation on quiet days), over the sunlit hemisphere (left) and the night hemisphere (right); (above) sunspot minimum at the equinox.
(below) The current-system at sunspot-minimum, in the northern summer.
(Chapman & Bartels, Geomagnetism, p.696.)

by 50 per cent or more. The currents are of less intensity by night than by day, corresponding to the reduced electron density at that time.

If we knew the systematic winds in the ionosphere, we could calculate from them and from the electron density distribution the *Sq* and *L* currents. This would fully explain the external part of the *Sq* and *L* fields at the earth's surface. Actually we do not yet know the winds, though radio and other studies are gradually giving information about them. The winds that produce *L* must be wholly tidal, caused by the same gravitational forces that produce the major part of the sea tides. The *Sq* winds are caused by both the tidal and thermal action of the sun.

1.5b. The electrical conductivity in the ionosphere

Progress towards such an inductive theory of the *Sq* and *L* variations has been made, and the dynamo origin of these changes is now accepted. The theory is much complicated by the anisotropic nature of the electrical conductivity of the ionosphere, caused by the presence of the main magnetic field. Charged particles can move freely along the field, but are forced to *spiral* round the lines of magnetic force. This reduces their mobility transverse to the field. Another complication is that the electromotive forces may include a vertical component. This cannot produce continuous current flow vertically, because of the insulating under-surface of the ionosphere. Instead it causes a vertical electric polarization, which modifies the dynamo-induced electric field. The result is that the equations of horizontal electric current flow have the form (Baker and Martyn, 1953):

$$i_{\mathrm{N}} = k_{\mathrm{N}}E_{\mathrm{N}} - k'E_{\mathrm{E}},$$

$$i_{\mathrm{E}} = k'E_{\mathrm{N}} + k_{\mathrm{E}}E_{\mathrm{E}}. \qquad (1.23)$$

Here the subscripts N and E refer to the geographical north and east; the k's are the tensor conductivity coefficients. They depend on the ionospheric height-distributions of $n_e(h)$ and $n_i(h)$, the electron and ion-densities, and also on the local magnetic inclination or dip I, and the intensity F. Thus (Chapman, 1956):

$$k_{\mathrm{N}} = k_0k_1/A, \quad k_{\mathrm{E}} = k_1(k_0S^2 + k_3C^2)/A, \quad k' = k_0k_2S/A, \qquad (1.24)$$

where

$$C = \cos I, \; S = \sin I, \; A = k_0S^2 + k_1C^2, \qquad (1.25)$$

and

$$k_0 = \frac{n_e e}{F}\left\{\frac{1}{X_e} + \frac{1}{X_i}\right\} + \frac{2n_- e}{FX_i} \qquad (1.26)$$

where n_- denotes the number density of negative ions; for simplicity these are supposed to have the same mass as the positive ions, whose number density is the sum of n_e and n_- (this is the condition of electric neutrality). Also the positive numbers X are given by

$$X_e = m_e\nu_e/eF, \quad X_i = m_i\nu_i/eF,$$

where ν_e and ν_i denote the collision frequencies and m_e, m_i the masses of the electrons and ions respectively, namely 9.1066×10^{-28} and 1.672×10^{-24} gm. Further

$$k_1 = \frac{n_e e}{F}\left\{\frac{1}{X_e + 1/X_e} + \frac{1}{X_i + 1/X_i}\right\} + \frac{2n_+ e}{F(X_i + 1/X_i)} \tag{1.27}$$

$$k_2 = \frac{n_e e}{F}\left\{\frac{1}{1 + X_e^2} - \frac{1}{1 + X_i^2}\right\} \tag{1.28}$$

and

$$k_3 = k_1 + k_2^2/k_1. \tag{1.29}$$

When $n_- = 0$,

$$k_3 = \frac{n_e e}{F}\frac{X_e + X_i}{1 + X_e X_i} = \frac{X_e X_i}{1 + X_e X_i} k_0 < k_0. \tag{1.30}$$

These conductivity coefficients are functions of height in the ionosphere, and so also may be the electric field components E_N, E_S. Calculations have been made for a model atmosphere for the latitude of White Sands, New Mexico, for the following distribution of n_e, taking $n = 0$; between the heights indicated, n_e is taken to increase linearly with height.

Below 80 km, $n_e = 0$;
At 80 km, $n_e = 1000/\text{cm}^3$;
At 90 km, $n_e = 10000/\text{cm}^3$;
At 100 km, $n_e = 100000/\text{cm}^3$;
From 150 km to 200 km, $n_e = 200000/\text{cm}^3$;
At 250 km, $n_e = 250000/\text{cm}^3$.

Two distributions of temperature were considered; the temperature and the mean molecular weight W of the air determine the air density distribution. In both temperature distributions, T was taken to be 205°K at 80 km, and 220° at 90 km; above that level a constant gradient of either 3°/km or 6°/km was adopted. Also W was taken to be 29 from 80 to 100 km, 25 at 150 km, and 19 at 300 km, varying uniformly between these levels.

Figure 5 shows the resulting height distribution of the coefficients k_1 to k_3 (and also that of the assumed n_e), for the hotter atmosphere above described. Clearly most of the conductivity is provided by the air between 90 and 150 km. At the magnetic poles, where $I = 90°$,

$$k_N = k_E = k_1, \quad k' = k_2.$$

At the magnetic equator, where $I = 0$,

$$k_N = k_0, \quad k_E = k_3, \quad k' = 0.$$

Near the magnetic equator the daytime electromotive force is approximately eastward, and k_3 is large. This causes an enhancement of the Sq and L currents in that region, by a factor of 2 or more; this feature is not shown in Fig. 4, but is indicated in the magnetic records of the Huancayo (Peru) observatory. The concentration of enhanced current in a narrow strip along the magnetic equator (or elsewhere) is called an *electrojet*, by analogy with the jet streams in the stratosphere. The equatorial electrojets of the *Sq* and *L* current systems are special features of the transient magnetic variations in this region, which also show peculiarities of electron distribution, of much interest for investigators of the ionosphere.

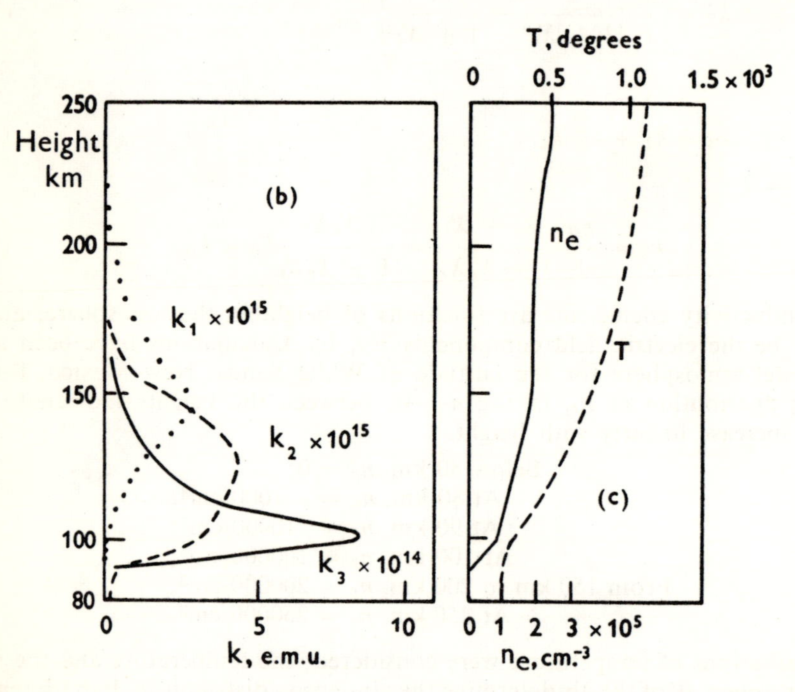

Fig. 1.5. The electrical conductivities k_1, k_2, k_3 in e.m.u. as functions of height in the ionosphere: for the electron density distribution n_e and temperature distribution T shown in the right-hand diagram. (Chapman, 1956; redrawn in Ratcliffe, 1960, p.395.)

Because of the importance of the magnetic dip I for geomagnetic and ionospheric phenomena near the dip equator ($I = 0$), it is found convenient to define a (magnetic) *dip* latitude Φ' in terms of the value I of the dip, at points not many degrees from the dip equator. The definition is based on the dipole relation (6), and is thus written:

$$\tan \Phi' = \tfrac{1}{2} \tan I. \qquad (1.31)$$

Near the magnetic poles, where k' exceeds k_N and k_E, the electric current flow is nearly perpendicular to the applied emf; for example, in that region a southward emf (E_N negative) would impel a nearly westward current.

Current diagrams such as those in Fig. 4 indicate the integrated current flow throughout the thickness of the ionosphere.

1.5c. Solar flares, short-wave radio fadeouts and the Sq currents

The association of solar flares with disturbances of radio communication is clearly established. Friedman (1960) and his colleagues have shown by rocket measurements that solar X-rays, which can penetrate to *D*-region levels, are enhanced during flares. Hence the *D* layer at such times becomes more ionized, more absorbent for radio waves, and more electrically conducting.

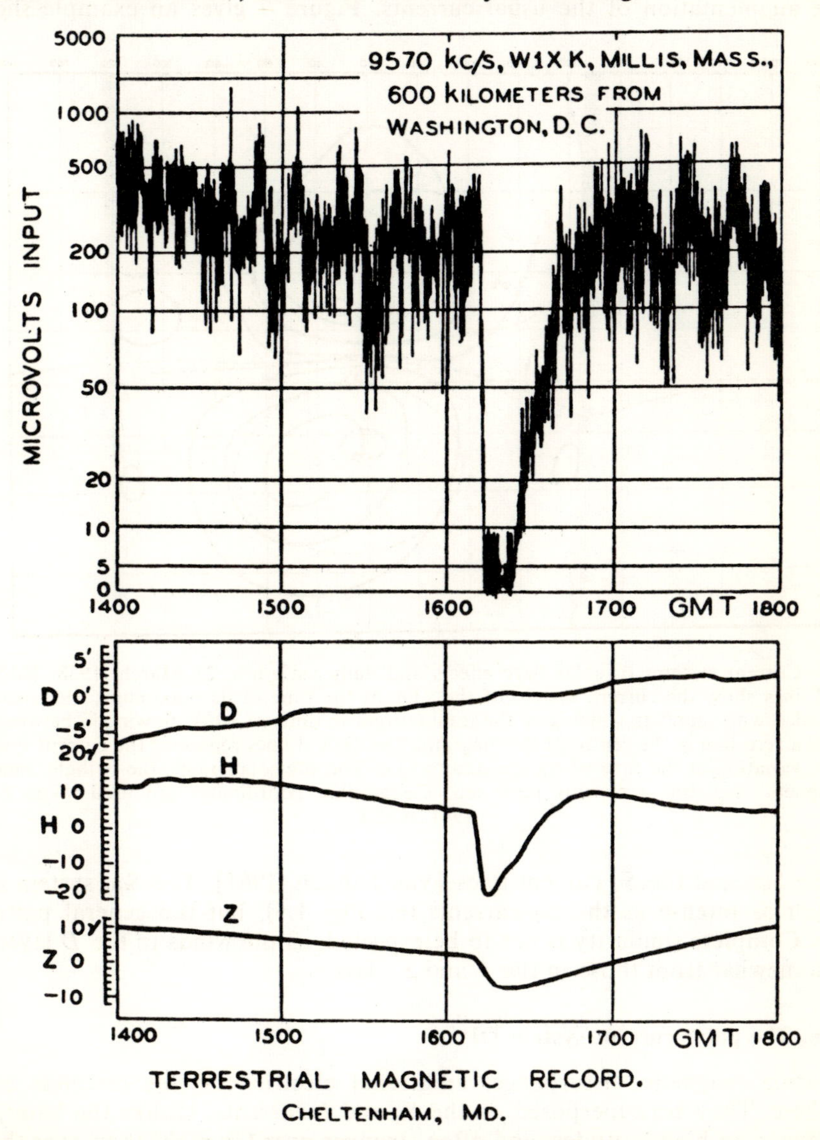

Fig. 1.6. The radio fade-out (low signal intensity) and magnetic disturbance on November 26, 1936 (Dellinger 1937; also Geomagnetism, p.348.)

Winds in the *D* region must at all times induce electromotive forces there; usually these are ineffective because of the low electrical conductivity. But during the brief life of flares the increased conductivity enables electric currents to flow. Their magnetic field shows as a fleeting disturbance in the magnetic records (Fig. 6). Study of these disturbances shows that the current pattern resembles that of the *Sq* system; the additional currents and their field are denoted by *Sqa*, the *a* signifying the augmentation of the usual currents. Figure 7 gives an example showing

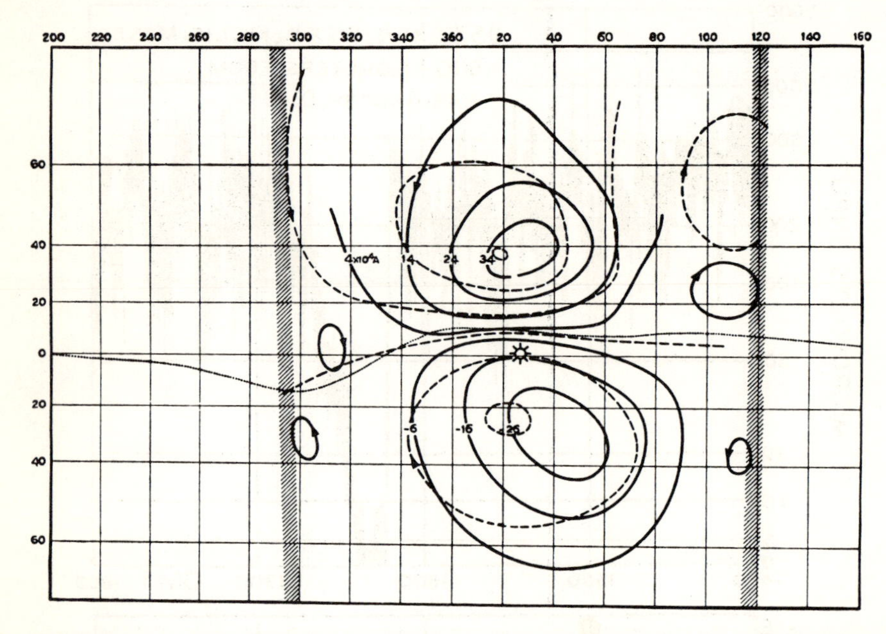

Fig. 1.7. Current systems of solar flare effects and daily variation, 23 March, 1958, 1015 U.T. The full lines show the current system of the s.f.e. at the time of its maximum; the streamlines are provided with numbers which give the total current in units of 10^4A, flowing in between these lines and a zero line in the centre of the diagram. The dashed lines represent the current system of the daily variation at the time of the maximum s.f.e. The sub-solar point, the twilight zones and the magnetic equator are also indicated. Geographic coordinates are used. (Van Sabben 1961, p.32.)

both the *Sqa* and the *Sq* current lines [Van Sabben, 1961]. The *Sqa* system is less than half as intense as the *Sq* currents (cf. Fig. 1.5), but the general pattern is similar. Complete similarity is not to be expected, as the winds in the *D* layer may differ somewhat from those in the *E* and *F*1 layers.

1.6. The polar current system DP

During magnetic disturbance, additional electric currents circulate in the ionosphere. They are superposed on the *Sq* and *L* currents. Unlike the latter, they are strongest in high latitudes, and often stronger over the night than over the day hemisphere. They are denoted by *DP*—the *D* signifies disturbance, and the *P*, polar, referring to the regions whence they are driven, and where they are strongest. They are especially concentrated along the auroral zones; hence the currents there are called the *auroral electrojets.*

Figure 8a illustrates a rather typical form of the northern hemisphere *DP* currents, as seen from above B, the boreal magnetic axis pole. The arrow towards the sun indicates the noon meridian. This current system corresponds to a magnetic "bay", in which the magnetic curves deviate for a short time from their usual course. Weak bays may affect mainly temperate and high latitudes, strong ones may affect all latitudes. Bays occur with special strength during magnetic storms. They may last for an hour or two, sometimes even longer, with fluctuating intensity.

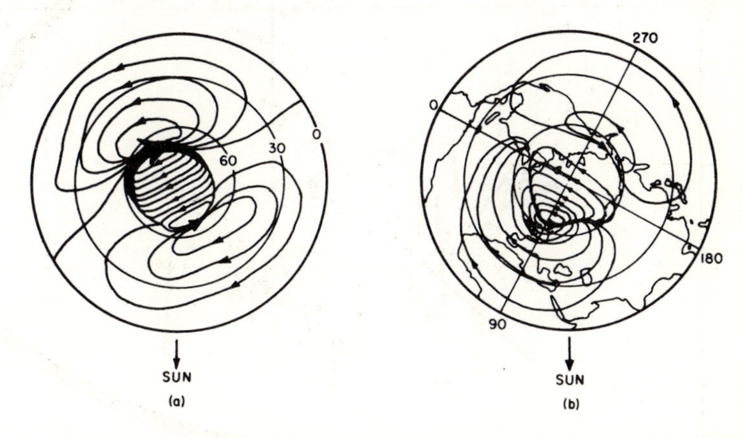

*Fig. 1.8*a. The current system of an average bay disturbance, viewed from above the geomagnetic pole. (Silsbee and Vestine 1942, p.206)

*Fig. 1.8*b. A tentative current system for the abnormally large *DP* sub-storm at 9h GMT on July 15, 1959. (Akasofu and Chapman, 1960, p. 103.)

Several may occur during one magnetic storm (in which there is a sudden commencement, and one main minimum of *H* due to the ring current); each is called a *DP* substorm. The stronger auroral electrojet, at such times, may carry a current of order a million amperes. The return current flows mainly over the polar cap, but partly also *between* the two auroral zones.

Figure 8b shows a rather unusual *DP* current system, derived from the data for the storm of July 15, 1959, at 9h U.T. It is based on the current arrows inferred from the magnetic records at many observatories at that time, as shown in Fig. 9. The sunlit half of the globe, bounded by the shadow line shown, is in the center of the diagram. The current system as drawn is tentative, and the evidence is clearest for the westward return current flowing south of the eastward auroral electrojet on the day side of the earth. Unfortunately many of the polar records were so disturbed as to prevent their being reliably measured. In some instances the trace went off the paper, even for the insensitive recorders. In other cases the fluctuations were so rapid and extreme that the photographic trace was too weakly shown to be scaled (Akasofu and Chapman, 1960). On this occasion the auroral electrojet may have carried a current of order 10 million amperes, far more than is usual, even in magnetic storms.

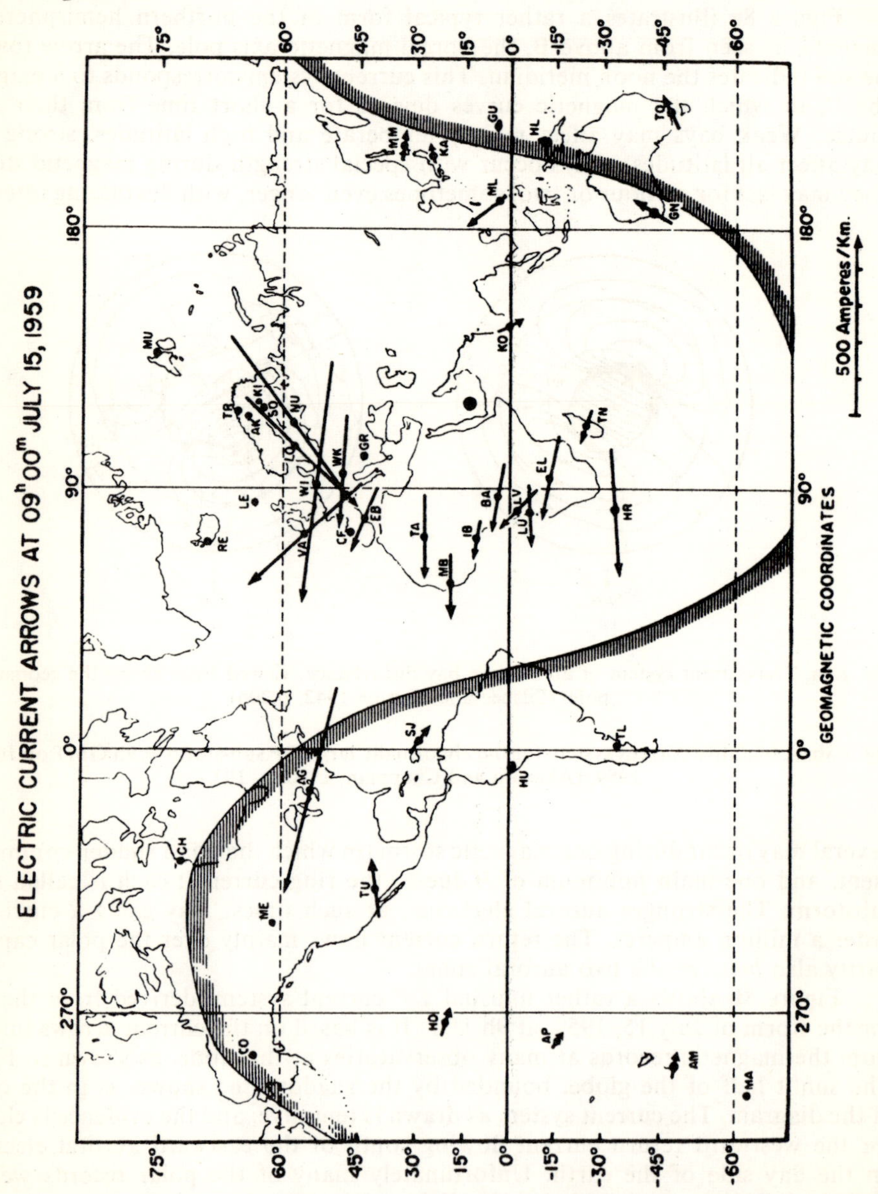

Fig. 1.9. Equivalent electric current vectors at 9h on July 15, 1959, corresponding to Fig. 1.8b. Notice the strong westward current over Europe and Africa. The sub-solar point and twilight sectors are indicated. (Akasofu and Chapman, 1960, p. 101.)

The *DP* currents are believed to be one of the many effects caused by the impact of a solar stream or cloud of ionized gas upon the earth's magnetic field. Some of the gas finds its way into the polar atmosphere, and causes the luminescence of the aurora. This entry of electrons and protons probably energizes the auroral electrojets; they take their return path in a way partly determined by the distribution of ionization over the earth. A dynamo theory seems inadequate to explain the *DP* currents, though dynamo effects may play some part in them.

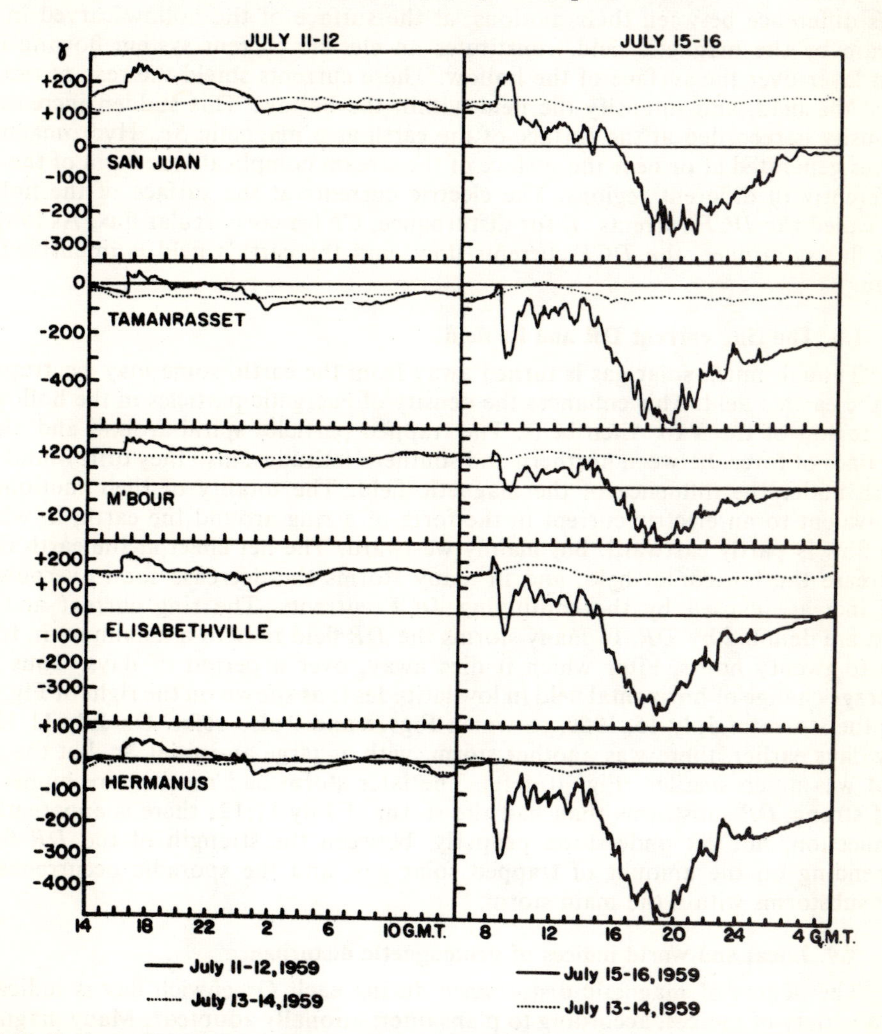

Fig. 1.10. Horizontal component magnetograms for 24 hour intervals beginning just before the storms of July 11, 12, 1959 and July 15, 16, 1959, from five stations in relatively low latitudes. The stations are well distributed in longitude. The dotted lines show the records for the same 24 hours of Greenwich time on July 13, 14, 1959 (between the two storms) when the magnetic activity was low. (Akasofu and Chapman, 1960, p. 96.)

1.7. The sudden commencement current DCF

Many magnetic storms, and practically all great storms, have a sudden commencement (*Sc*), simultaneous all over the earth to within a minute or less. The sudden commencement is ascribed to the impact of the solar ionized gas on the outer part of the geomagnetic field, at a distance of several earth radii—20 to 40 thousand miles from the center of the earth. The particles are turned partly back, partly sideways, and the deflection is different for the electrons and protons. This difference between their motions, at the surface of the hollow carved in the stream by the magnetic field, constitutes an electric current system flowing in a thin layer over the surface of the hollow. These currents shield the rear of the gas from the field, and intensify the field within the hollow. This sudden increase of intensity is recorded at the surface of the earth as a magnetic *Sc*. Hydromagnetic waves generated at or near the surface of the stream complicate the form of the *Sc*, differently in different regions. The electric currents at the surface of the hollow are called the *DCF* currents, *D* for disturbance, *CF* for corpuscular flux. As long as this flux continues, the *DCF* currents flow, and the earth's field is intensified by them.

1.8. The ring current DR and its field

Though much solar gas is turned away from the earth, some may be trapped by the earth's field. This enhances the density of energetic particles in the hollow—the region of the Van Allen belts. The trapped particles spiral around and along the lines of force, between northern and southern latitudes; also they drift round the earth under the influence of the magnetic field. The totality of their motions is equivalent to an electric current in the form of a ring around the earth, in which the flow is partly eastward, but mainly westward. The net effect at the earth is to decrease the horizontal field, and in many storms the decrease soon overpowers the increase caused by the continuing *DCF* currents. The ring current and its field are denoted by *DR*. In many storms the *DR* field rises to a maximum in from ten to twenty hours, after which it dies away, over a period of days. Thus the average change of horizontal field in low latitudes is as shown on the right of Fig. 10, for the storm of July 15, 1959, to which Figs. 8 and 9 also refer. On July 11/12, a few days earlier, there was another storm, with as large an initial *Sc*, but the *DR* field was much smaller [Fig. 10 left]. The later storm had much more numerous and strong *DP* substorms than had the storm of July 11/12; there is apparently a connection, not yet understood properly, between the strength of the *DR* field, depending on the amount of trapped solar gas, and the sporadic occurrence of *DP* substorms within the main storm.

1.9. Local and world indices of geomagnetic disturbance

The degree of magnetic disturbance during each Greenwich day is indicated by a variety of indices, according to plans internationally adopted*. Many magnetic

* The plans are under the auspices and supervision of the Committee on Characterization of Magnetic Disturbances of IAGA (International Association of Geomagnetism and Aeronomy), a part of IUGG (International Union of Geodesy and Geophysics); the present Chairman is J. Bartels, and the center at De Bilt is in charge of J. Veldkamp.

observatories throughout the world take part in these plans. They assign *local* indices of magnetic activity. These are sent to an international center at De Bilt, The Netherlands. This is linked with a Permanent Service of Geomagnetic Indices at Göttingen, Germany.† There the local indices are used as the basis for world indices.

Each participating observatory assigns to each Greenwich day its own "daily character figure" C. This is one of the numbers 0, 1 or 2, to indicate ascending degrees of disturbance of its records. This local daily index is assigned by simple inspection of the records. Naturally there are many border-line cases where the choice between 0 and 1, or between 1 and 2, is difficult; but the choice is made.

The values of C from all observatories for each day are averaged to one decimal place. This gives Ci, the "international daily character figure". It provides a 21-fold classification of Greenwich days, from 0.0 on days of extreme calm, to 2.0 on days that are magnetically highly disturbed.

By a more detailed and precise method a local index K is assigned to each 3-hourly interval of Greenwich time, 0–3, 3–6, . . .; K is one of the integers 0 to 9, giving a 10-fold classification of these intervals. The records of the three elements are examined, and for each element an estimate is made of the range r during the interval, allowing for the part of the change caused by Sq and L, and, when necessary, by Sqa and DR, in the recovery phase of the ring current field. Such allowance calls for experience and judgement. The largest of the three ranges is taken as the basis for K; the element concerned may vary from interval to interval, or from one observatory to another.

For each observatory a table is assigned, giving limits of r corresponding to each of the ten values of K. The lower limit of r for $K = 9$ may have any of the values 300, 350, 500, 600, 750, 1000, 1200, 1500 and 2000γ. The first applies to very low latitude observatories (outside the belt of the equatorial electrojet; §1.5b). The last applies to the most disturbed stations, in the auroral zone. The limit 500 refers to stations in about geomagnetic latitude 50°. The table for this latitude is as follows:

Range of																				
r:	0		5		10		20		40		70		120		200		330		500	
K:		0		1		2		3		4		5		6		7		8		9

For other latitudes, the lower values of the range limits of r are scaled up or down proportionately to the lower limit for $K = 9$.

The C and K figures for some observatories appear currently in their own or other publications. A volume containing the complete set for all participating observatories, together with much other data, including the world indices, is published annually as an issue of IAGA Bulletin 12; the latest volume is 12*l*, for 1957.

† Addresses: C + K Centre, Kon. Ned. Meteor. Inst., De Bilt, The Netherlands; and Geophysikalisches Institut, 180 Herzberger Landstr., (34) Göttingen, Germany.

These volumes, formerly issued as IATME* Bulletins 12, are now distributed by the North Holland Publishing Co., Amsterdam, from whom information as to the availability of back issues can be obtained.†

The values of *K* for any one observatory show statistically a daily variation depending on its local time. To obtain a world index this local influence must be eliminated. Corrected or standardized values (each denoted by *Ks*) of *K* are prepared at the Permanent Service at Göttingen, for each of 12 selected observatories, northern and southern, lying between dipole latitudes 48° and 63°. The mean of the 12 values of *Ks* for these 12 observatories is denoted by *Kp*; it is called the planetary 3-hour index and is expressed in a scale of thirds,

$$0_0,\ 0_+,\ 1_-,\ 1_0,\ 1_+,\ 2_-,\ 2_0,\ \ldots,\ 9_-,\ 9_0.$$

Thus it is a 28-fold classification.

By cooperation between De Bilt and Göttingen, it is possible to issue monthly tables of *Kp* before the end of the following month. The values are also published in the *F*-series: Ionospheric Data of the Radio Propagation Physics Division, National Bureau of Standards, Boulder, Colorado: in the Journal of Geophysical Research; and elsewhere. Quarterly Bulletins containing the *Kp* figures and other magnetic data are issued from De Bilt.

Along with the values of *Kp*, values of two *daily* world indices *Ap* and *Cp* are given. These are based on the sum of the eight values of *Kp* for each day. The values of *Cp* range, like those of *Ci*, from 0.0 to 2.0 at intervals of 0.1. A table of range limits for the daily sums of *Kp* has been formed, such that the frequency distributions of *Ci* and *Cp* for the ten years 1940–9 are nearly identical. The indices *Ci* and *Cp* are approximately *logarithmically* related to the ranges *r* of the magnetic elements during disturbance (corrected for *Sq*, *L* and *Sc*). The indices *Ap* are more nearly *linearly* related to these ranges.

The planetary index *Kp* is designed to measure solar particle radiation by its magnetic effects, specifically to meet the needs of research workers in the ionospheric field.

The *K* tables for the individual observatories, in the IAGA Bulletins 12, give a very detailed indication of the incidence, in time and place, of the magnetic effects of the solar eruptions. The *Kp*, *Cp*, *Ap* and *Ci* tables give a concise indication of the overall world effects [Bartels, 1957]. These tables are now available from the time of the 2nd International Polar Year (1932/3) to date. The tables of *Ci* go back still further. Details regarding these publications are given in Bulletin 12*l* (pp. iv and 268); see also IGY Annals, vol. 4, pp. 227–236, where also, on pp. 220–226, an account is given of the 1/4-hourly indices *Q* to be issued, covering the period of the IGY. These will permit studies of correlations between magnetic disturbance and other phenomena, in much more time-detail than is possible with the daily and 3-hourly indices.

* IATME signifies International Association of Terrestrial Magnetism and Electricity, the former name of IAGA.

† This publication and the work of the two centers for geomagnetic indices, at De Bilt and Göttingen, are sponsored by ICSU (International Council of Scientific Unions) with financial support from UNESCO.

Finally, note in this connection the publication of diagrams (Bartels' "musical scales") graphically illustrating the sequence of the *Kp* values. These are arranged in rows so as to show the 27-day recurrence tendency of magnetic disturbance, arising from the solar rotation. The monthly diagrams are issued monthly along with the *Kp* tables, in time to be helpful in making 27-day recurrence forecasts.

The monthly diagrams issued during the IGY have been combined into a diagram for the whole IGY period, and since then annual diagrams have been issued (see Bulletins 12 for further information).

For longer-term studies of the intensity of geomagnetic disturbance, Bartels (1932) introduced an index u called the inter-diurnal variability; see Fig. 1.2. For a particular station P the inter-diurnal variability U for any day is the numerical difference between the mean values of the horizontal intensity H for that day and for the preceding day, taken without regard to sign. Its average amount depends on the (magnetic) dipole latitude or (magnetic) dipole north-polar distance θ of P. An equatorial value u of the inter-diurnal variability is obtained by taking the mean of $U/\sin\theta\cos\psi$ for a number of observatories, where ψ denotes the angle between **H** at P, and the great circle that passes through P and the magnetic-axis poles. The division by $\sin\theta\cos\psi$ is made in order to base the index u on the component of **H** parallel to the dipole axis. This equatorial value u, expressed in units of 10γ, is taken as a measure of the magnetic activity. It is not appropriate as a measure for a single day, but its monthly or annual mean values measure the average amount of one aspect of magnetic disturbance during the interval; for further details concerning this measure see *Geomagnetism* (Chapman and Bartels, 1940, Chapter 11).

1.10. The solar rotation

Carrington, an English amateur astronomer, published an important volume (1863) *Observations of the Spots on the Sun*, based on his work from 1853 to 1861. From his observations of the motions of the spots across the sun's disc he determined the elements of the sun's axis, and the mean angular velocity of the sun in different latitudes; it is greatest in the lowest latitudes. He adopted a mean angular velocity, in order to define a prime (rotating) meridian of reference of heliographic longitude of markings seen on the sun. His adopted sidereal rotation period is 25.38 mean solar days. This corresponds to a mean *synodic* rotation period of 27.275 days, that is, the period of rotation relative to the earth, or the time between successive passages through the central meridian as seen from the earth. Carrington's meridian passed through the ascending node at the epoch 1854.0. Its successive rotations since 1853 November 9 are numbered serially.

The synodic periods in days, derived from spots in different latitudes, are as follows (Maunder, 1922):

Latitude on the sun:	0°	5°	10°	15°	20°	25°	30°	35°
Synodic period, days:	26.4	26.5	26.6	26.8	27.1	27.5	27.8	28.7

However, the rotation periods determined from different spots in any one narrow belt of latitude cover a range at least as wide as that of the average periods shown in this Table.

The sun's equatorial plane is inclined at 7.3° to the ecliptic or plane of the earth's orbit. The two planes intersect along a line through which the earth passes on or about June 6 and December 8; thus on these days the earth's heliographic latitude is zero. Between June 6 and December 8 the earth's heliographic latitude is positive (northern), and its maximum value 7.3° is attained on about September 8. On about March 6 it attains its southern maximum. These dates refer to the year 1934; they change slightly from year to year, and are given in various ephemerides for each year.

The sun's geographic latitude, or declination, is zero at the spring and autumnal equinoxes, March 21 and September 23; its northern and southern maxima. ($\pm$ 23.5°) are attained at the solstices June 22 and December 22.

1.11. Measures of solar activity

Rudolf Wolf of Berne, first director of the Swiss Federal Observatory at Zürich, sought and studied all available data for sunspots from 1610 (when Galileo first saw them through his telescope, and thereby discovered the sun's rotation). Wolf defined a relative sunspot number r for each day, by the formula

$$r = k(f + 10g)$$

Here g denotes the number of sunspot groups (however many their component spots) and f the total number of spots in all the visible groups. Values of f and g are sent to Zürich by various observatories, and k is a factor there applied to the sum $f + 10g$, depending on the estimated efficiency of the observer and his telescope. This work has been continued by Wolf's successors Wolfer, Brunner and Waldmeier, and the Zürich observatory became and has continued to be a main center for sunspot data. At present over forty observatories and private astronomers contribute to the Zürich relative sunspot numbers. Thus they ensure a complete day by day record, despite gaps due to clouds or other causes in the individual series of observations. Long series of annual and monthly mean sunspot numbers are available (Newton, 1958; Waldmeier, 1961).

The *areas* and heliographic positions of sunspots based on photographic records at Greenwich and cooperating observatories are annually published in the *Greenwich Photoheliographic Results*. An important volume *Sunspots and Geomagnetic Storm Data* 1874–1954, *Greenwich* (1955) summarizes the Greenwich records of the two related sets of phenomena over an interval of 80 years.

Under the auspices of the International Astronomical Union the Zürich Observatory publishes a *Quarterly Bulletin on Solar Activity* (with financial support from UNESCO). It gives relative sunspot numbers and areas for each day, and tables and drawings illustrating many other characteristics of solar activity.

1.12. Sunspot latitudes and magnetism

Carrington recognized that the solar cycle of sunspot activity is accompanied by a variation in the mean latitudes as well as in the number of spots. His work was confirmed and extended by Spörer. Maunder (1922) illustrated the latitude variation in a diagram of the kind shown in Fig. 1.11, as brought up to a later date in the Greenwich volume of sunspot and magnetic storm observations for 1874–1954, above mentioned. At the beginning of each cycle, reckoned from sunspot minimum, spots of the new cycle begin to appear in latitudes 30° to 35°, north and south of the solar equator. Thereafter the latitude of appearance moves irregularly towards the equator, to reach a mean value of about 7° near the sunspot minimum. Near the minimum there are often spots of the old cycle in low latitudes at the same time as spots of the new cycle in the higher latitudes. Spots or groups of spots appear at irregular and unpredictable intervals; any one spot or group lasts at most for a few weeks.

Hale (1908), studying the spectrum of sunspots, showed that spots have strong magnetic fields, as revealed by the Zeeman effect on the spectral lines. In the typical bipolar spot group the magnetic polarity of the leader spot is opposite to that of the follower spot; on opposite sides of the sun's equator the leader spots have opposite polarities. At sunspot minimum there is a reversal of the polarity of the spots of the new cycle. Thus the sunspot cycle has a period of about 22 instead of 11 years. In recent years the technique of solar magnetic observation has been greatly improved (Babcock, 1953). The general magnetic field of the sun has been found to suffer reversal near sunspot maximum. Various theories of sunspot and general solar magnetism have been proposed, but the subject is still obscure (Alfvén, 1956; Allen 1960; Babcock, 1961; Steenbeck, 1961; Walén, 1949).

1.13. The 27-day recurrence tendency in geomagnetic disturbance

The general relationship that connects the sunspot cycle with various indices of geomagnetic variation (Fig. 1.2), is most apparent in the annual means, and scarcely at all evident in the daily values. Besides this relationship there is another, now known as the 27-day recurrence tendency. Notable departures of

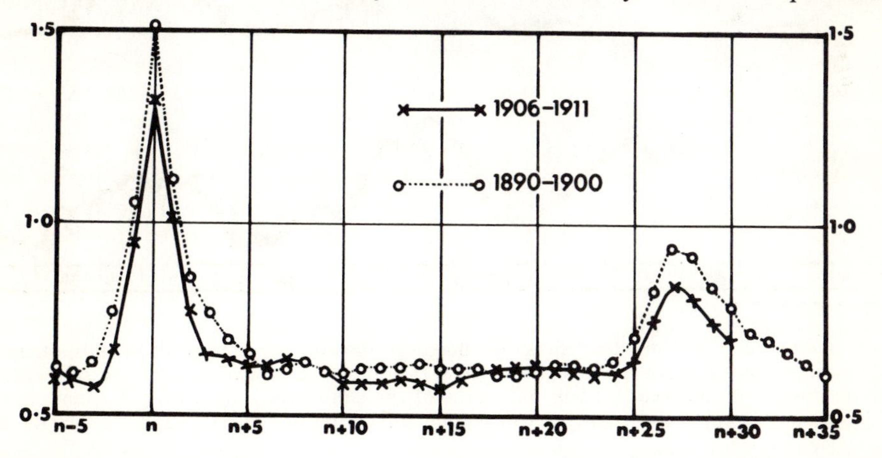

Fig. 1.12. The 27-day recurrence phenomenon in the daily magnetic character-figures, as shown by Chree's daily character-figures assigned on the basis of the Kew records 1890–1900, and by the international magnetic character-figures, 1906–11. The curves indicate the average daily magnetic character-figures for days preceding and following a set of selected magnetically disturbed days (*n*). (Chree 1908; also Geomagnetism, p,397.)

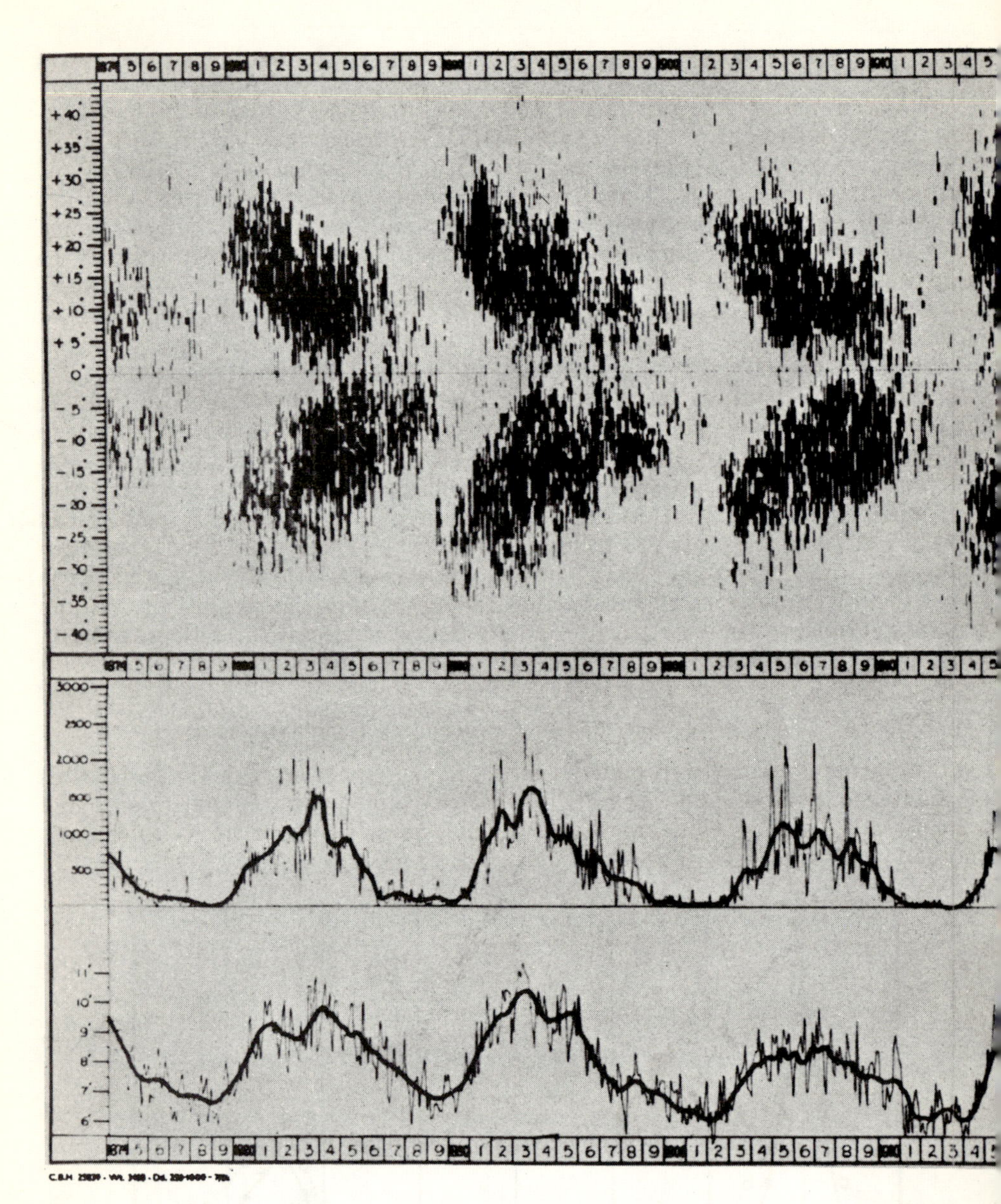

Fig. 1.11. (Above) The Maunder "Butterfly" diagram, showing the occurrence of sunspots in time and extent in solar latitude, from 1874 to 1954. (Middle) The spotted area of the sun, corrected for foreshortening, and expressed in millionths of the sun's visible disc, over the same interval; the lightly drawn curve shows the mean for each solar rotation (27 days) and the heavily drawn

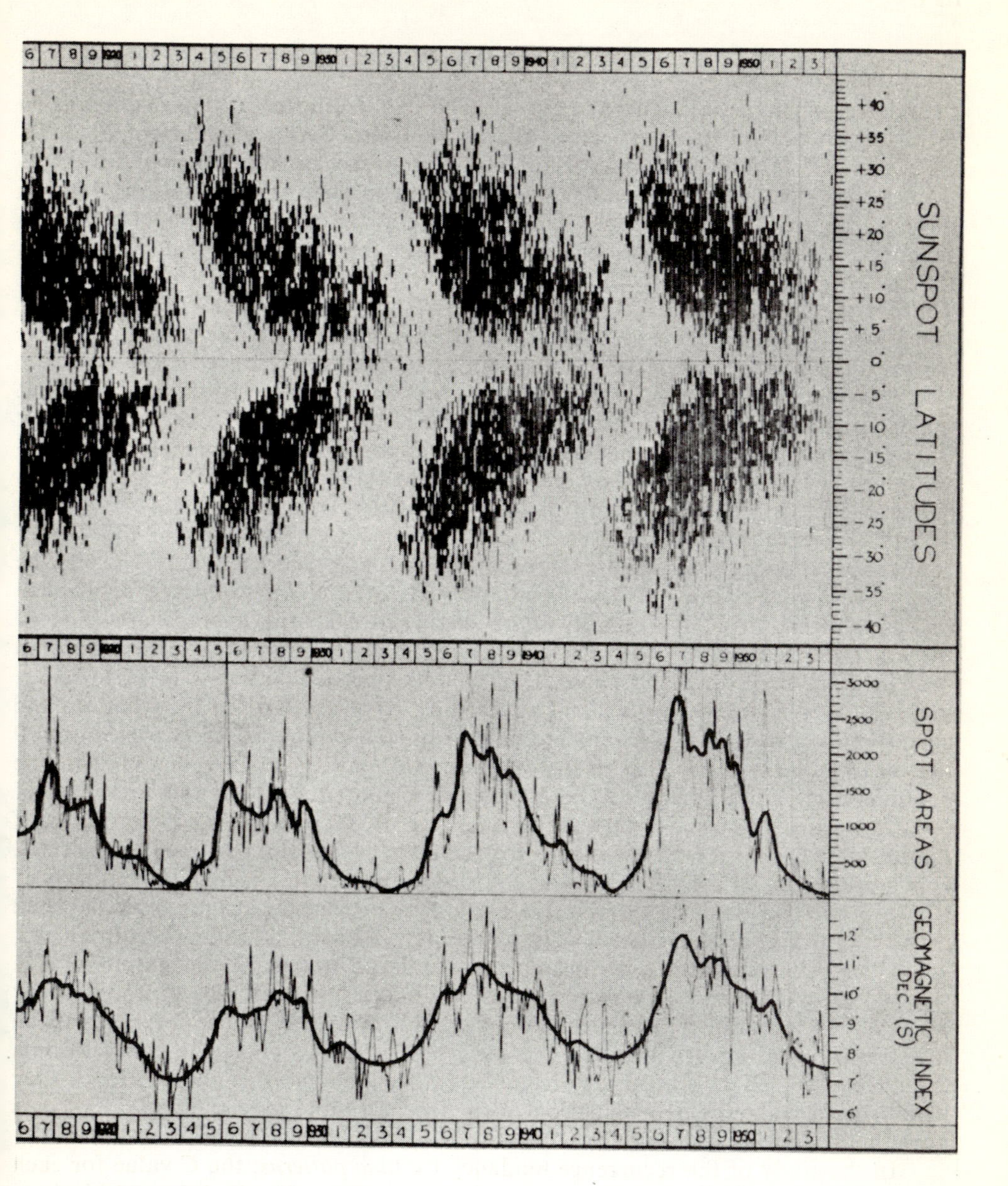

curve, the annual mean area; This curve shows clearly the epochs of sunspot maximum and minimum, for comparison with the upper diagram. (Below) The daily range of magnetic declination at Greenwich; means for each solar rotation, and annual means, in minutes of arc. (Greenwich 1955).

the degree of magnetic disturbance on a given day, from the current average level of disturbance, tend to be repeated, usually in diminished degree, about 27 or 28 days later. For a time this was interpreted by some as an indication of lunar influence; but this view is now abandoned, as is natural because of the very limited physical influence that the moon can exert upon the earth. The phenomenon is called the 27-day recurrency *tendency*, because it represents no true periodicity. Its cause is now ascribed to the rotation of the sun relative to the moving earth. The sun does not rotate as a rigid body, so there is no unique solar rotation period, but 27 days is a reasonable estimate of the average recurrence interval; it approximates to the period of relative rotation of the main sunspot zones.

The tendency can be studied with some objective precision, and without imposing pre-conceived estimates of the recurrence period, in two ways, (i) by time patterns, and (ii) by the method of superposed epochs. Leaders in the development of the two methods were (i) Maunder (1904, 1905, 1916) and Bartels (1932, 1934); (ii) Chree (1912, 1913, 1927) and Chree and Stagg (1927).

In the method of superposed epochs, as applied to the study of the 27-day recurrence tendency in magnetic disturbance, the daily character figures C may be used. These range from 0.0 to 2.0, at intervals of 0.1. To study the recurrence tendency, numerous days notably more disturbed than the average are chosen: these are the selected epochs. Their values of C are written in a vertical column headed 0: in preceding and succeeding columns headed . . ., -2, -1, 1, 2, . . . the values of C for preceding and succeeding days are written in order, in the corresponding rows. Then means are taken for each column. Clearly for column 0 the mean value is high. The means on either side fall to lower values, but in the columns headed 27, 28 and 29 the means rise again. (Fig. 12). If the sequences of days following the selected days are extended to 60, 90, 110 days, smaller peaks of diminishing size appear in the means for the days whose number from the selected epoch is around 54, 55, also around 81 to 83, and around 108–111: similarly if the rows are extended backwards. Likewise, if the selected days are *less* disturbed then the average, so that their C values are notably low, the same procedure leads, when the columnar means are graphed, to a large trough in the graph for the selected (0) days, and to smaller troughs at intervals or multiples of about 27 or 28 days. As the unit of tabulation is 1 day, the mean recurrence interval can be determined only approximately: the successive pulses, when the tabulation is extended over 110 days after the selected days, become more rounded. Hence they help little to make the determination of the interval more precise; but 27.3 days is a rough estimate of its value.

In the study of the recurrence tendency by *time patterns*, the C value for each day included in the pattern is represented by the greater or lesser degree of blackening of a small square area associated with the day: no blackening indicates $C = 0$; and $C = 2.0$ corresponds to complete blackening. The range from $C = 0$ to $C = 2.0$ may be divided into sections, with all the C values in each section indicated by the same degree of blackening. The days are arranged in rows of 27, from some chosen initial day; but to show continuity, a few days may be added before or after each row, duplicating what is shown for them on the row following or preceding. A true 27-day recurrence would be indicated by a series of squares blacker or lighter than the average, in a vertical column. A recurrence tendency with a shorter or longer interval would be shown by similar repetitions, except that the series of two or more squares concerned would not be in the same vertical column; they would slope downwards to the left or the right.

Figure 13 shows two time patterns, Fig. 13a (on the left) for the daily magnetic character figures *C*, Fig. 13b (on the right, similarly constructed) for the daily sunspot numbers *R*.

Each of the two patterns includes 11 years' data, for the sunspot cycle 1923–1933, beginning and ending at a sunspot minimum. The sunspot maximum is shown near the middle of Fig. 13b, though much less clearly than in graphs of the annual means (Fig. 1). The epoch of maximum *geomagnetic* disturbance, shown by Fig. 13a, occurs a year or two after sunspot maximum. During sunspot minimum years the lull in magnetic disturbance is far less pronounced than in sunspots. Fig. 13b shows some instances of the recurrence tendency for sunspots, corresponding to the return of long-lived sunspot groups to the side of the sun facing the earth, after one or even two rotations. Many sunspots, however, live less long, and are seen only during one passage across the visible disc. But Fig. 13b clearly shows that certain longitudes on the sun, corresponding to particular columns in Fig. 13b, remain intermittently active (as shown by sunspots) over periods of many months.

The series of recurrences of magnetic disturbance, represented by blackened columns extending over several rows, are longer in the lower than in the upper part of Fig. 13a; that is, there is a greater tendency to recurrence after than before sunspot maximum. Some such recurrences extend over several solar rotations, and then, after an inactive period, are renewed.

1.13a. Interpretation of the recurrence tendency in geomagnetic disturbance

This tendency is ascribed to the continuing emission by the sun, from particular areas, over periods of weeks or months, of streams of solar gas, in a fairly radial direction. The magnetic disturbance is supposed to result in some way from the impact of the gas upon the earth. Such disturbance could recur on later impacts, after one or more solar rotations, so long as the emission continues in approximately the same direction. The recurrence might end by the sun ceasing to emit from the region, or by a change of direction of the stream, or by the earth's motion in its orbit taking it out of the region swept by the stream.

Solar observers have not as yet identified the regions of such emission. At times they may be in sunspot areas, but they may persist after the spots have died away. It is possible (but not yet proved) that these regions, which Bartels has named *M* regions, may coincide with unipolar magnetic areas on the sun, not necessarily marked by a sunspot.

The conical angle of the stream emission, and its variation from one stream and time to another, are not known. But two lines of evidence suggest values sometimes ranging up to about 30°. One indication is the occurrence of magnetic disturbance when the earth is on one side of the sun's equatorial plane, and the active areas on the sun lie on the other side of this plane. The angular distance of the earth from the sunspot area judged likely to be affecting the earth may sometimes be as great as 10°, and the earth must then be within the semi-angle of the cone of emission, if the emission is symmetrical about a solar radius. The

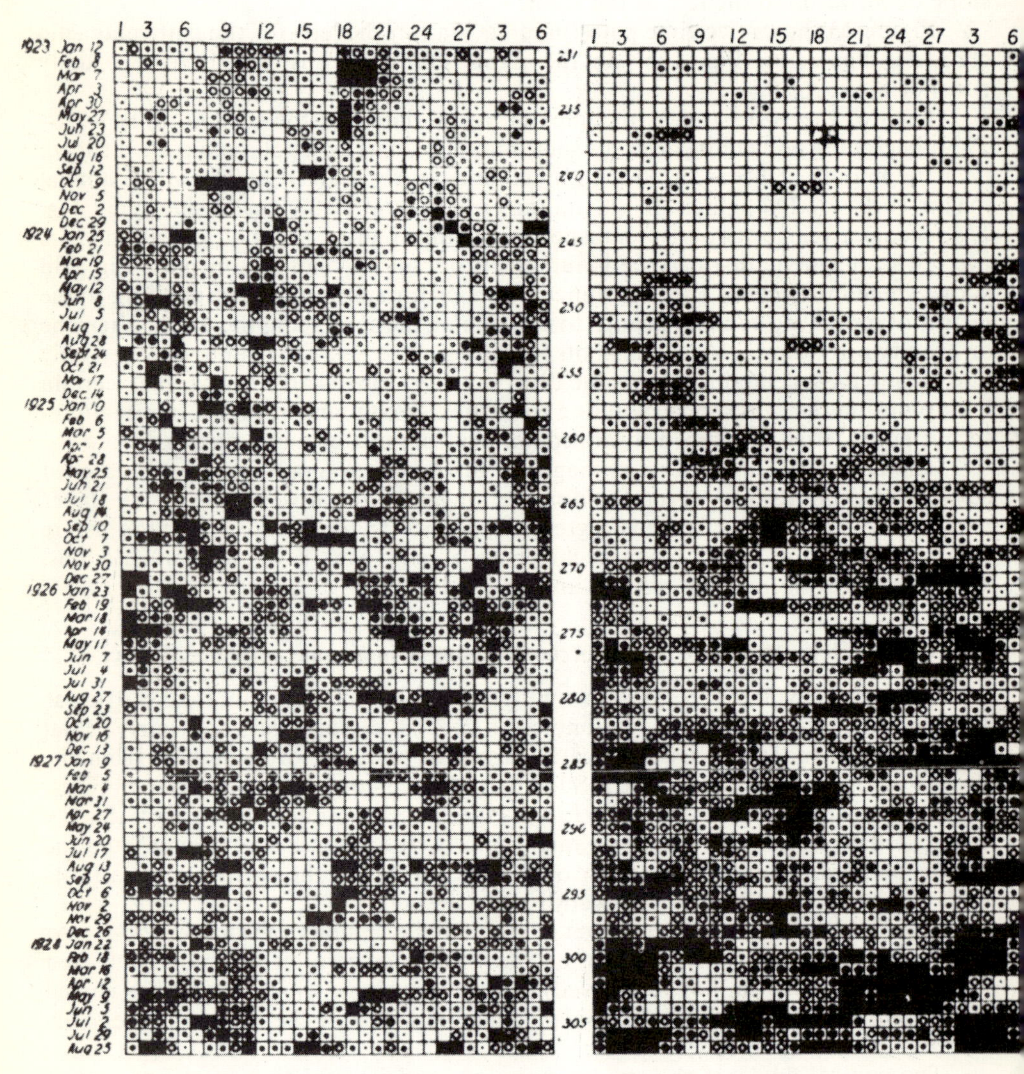

Fig. 1.13. Twenty-seven day recurrence time-patterns for geomagnetic activity (left) and for solar activity (right), 1923–1933. (Bartels, 1934; also Geomagnetism, p.411.)

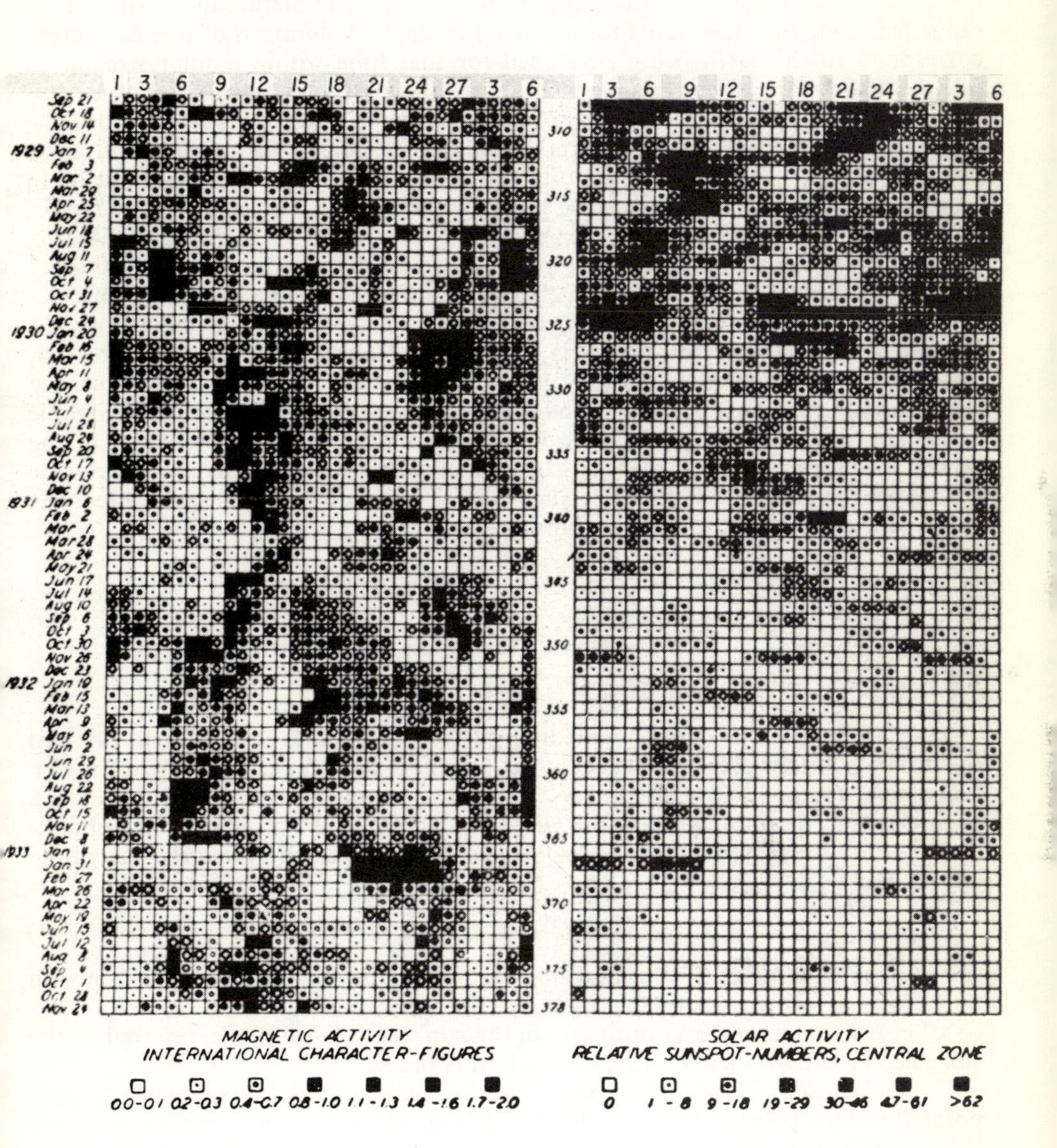
MAGNETIC ACTIVITY
INTERNATIONAL CHARACTER-FIGURES
00-01 0.2-0.3 0.4-0.7 0.8-1.0 1.1-1.3 1.4-1.6 1.7-2.0
SOLAR ACTIVITY
RELATIVE SUNSPOT-NUMBERS, CENTRAL ZONE
0 1-8 9-18 19-29 30-46 47-61 >62

other indication is given by the duration of geomagnetic disturbance, which may range from one or a few hours to one or a few days. A duration of one day might correspond to the earth being enveloped for that time within a solar stream. In one day the stream moves relative to the earth through a fraction 1/27.3 of a revolution of the stream with the sun, relative to the earth. This would indicate that the (complete) conical angle of the stream is at least 360°/27.3 or 13°: a longer duration within the stream would correspond to a larger angle. If the earth does not lie in an axial plane of the cone, this angle must be increased by an appropriate factor.

The angle subtended by the earth at the sun is 4.9×10^{-4} degree. At the earth's distance, a stream of conical angle 20° would have a diameter 40,000 times that of the earth. Thus relative to such a stream the earth is a very small object.

A solar meridian plane cuts the earth's orbit in a point whose speed along the orbit, relative to the earth's speed, is 400 km/sec. Hence it sweeps across the diameter of the earth in 32 seconds. This suggests that magnetic disturbance caused by such solar streams may at times have a rather sudden onset, affecting the whole earth almost simultaneously. This would correspond to a sharp boundary of the stream. However, as the stream is gaseous, its density is more likely to decrease gradually, outwards from its axis; such a stream would seem likely to affect the earth with a somewhat gradual beginning.

2. The Solar Plasma on its Way to the Earth

2.1. The path of a solar particle*

Let a ($= 7 \times 10^{10}$ cm) denote the sun's radius, and R ($= 1.5 \times 10^{13}$ cm) the distance of the earth from the sun. Let Ω denote the angular velocity of the sun, taken to correspond to Carrington's sidereal rotation period of 25.4 days (or a mean synodic rotation period of 27.3 days). Then $\Omega a = 2.0 \times 10^5$ cm/sec approximately. Let Ω' denote the mean angular velocity of the earth in its orbit round the sun. Because

$$R/a = 214, \ (R/a)^{1/2} = 14.6, \ \Omega/\Omega' = 14.4, \tag{2.1}$$

the reciprocals of these quantities can in many cases be ignored in comparison with unity.

Let r, θ denote polar coordinates in the sun's equatorial plane, referred to the sun's center O and a fixed initial radius in this plane.

First consider emission of particles in the plane of the sun's equator from a point P in that plane, situated in heliographic longitude λ, referred to a standard solar meridian which at time $t = 0$ has the azimuth $\theta = \Theta_0$. Then at time t the azimuth of P relative to the fixed radius in the equatorial plane is Θ, given by

$$\Theta = \Theta_0 + \lambda + \Omega t. \tag{2.2}$$

Suppose that the particles are emitted with constant radial and transverse velocity components v_0 and $\omega_0 a$, and that after emission the force (if any) on the

* This section reproduces, corrects and extends earlier results (Chapman 1929).

particles is radial. Then their angular momentum about O is constant, so that the transverse velocity component of a particle, in the equatorial plane of the sun, at distance r from O, is ωr, where

$$\omega r^2 = \omega_0 a^2. \tag{2.3}$$

The radial velocity component at distance r will be some function $f(r)$ of r, depending on the nature of the radial force.

Let r, θ now denote the polar coordinates (in the equatorial plane of the sun), at time t, of the particle emitted from P at time t_0. At time t_0 we have $r = a$, $\theta = \Theta$, $\dot{r} = v_0$, $\dot{\theta} = \omega_0$, where the superposed dot implies differentiation with respect to t. At time t we have $\dot{r} = v = f(r)$, $\dot{\theta} = \omega = \omega_0 a^2/r^2$. The geometrical equation of the path is obtained by eliminating the time from the last two equations, giving

$$d\theta/dr = \dot{\theta}/\dot{r} = \omega_0 a^2/r^2 f(r),$$

or, on integration,

$$\theta = \Theta_0 + \lambda + \Omega t_0 + \omega_0 a^2 \int_a^r dr/r^2 f(r). \tag{2.4}$$

Thus θ is a function of λ, t_0, ω_0, and r, but the *form* of the path is independent of Θ_0, λ, t_0; these merely determine the orientation of the path as a whole, in the sun's equatorial plane.

The inclination ϕ of the path to the radius, at any distance r, is given by

$$\tan\phi = r\, d\theta/dr = \omega_0 a^2/r f(r) = \omega_0 a^2/vr. \tag{2.5}$$

The time t of arrival at the distance r is given by

$$t = t_0 + \int_a^r dr/f(r); \tag{2.6}$$

the second term gives the duration of travel to this distance. From (6) r may be obtained as a function of t, and then (4) gives θ as a function of t.

Let Δt denote the time of passage to the earth's distance R; then

$$\Delta t = \int_a^R dr/f(r) \tag{2.7}$$

The value ϕ' of ϕ at this distance is given by

$$\tan\phi' = \omega_0 a^2/vR. \tag{2.8}$$

The angle $\Delta\theta$ subtended at O by the part of the path P to the distance r is given by

$$\Delta\theta = \omega_0 a^2 \int_a^R dr/r^2 f(r). \tag{2.9}$$

2.1a. Special case; no radial force

If the particles are subject to no radial force after emission, $f(r) = v_0$ always, and equations (4), (7) and (9) take the special forms

$$\theta = \Theta_0 + \lambda + \Omega t_0 + \omega_0 a/v_0, \tag{2.4a}$$

$$\Delta t = R/v_0, \tag{2.7a}$$

$$\Delta\theta = \omega_0 a/v_0 = \omega_0 a \Delta t/R, \tag{2.9a}$$

where factors $1 - a/R$ have been omitted; cf. (1).

2.1b. Special case: the Milne Doppler force

Milne (1926) showed that in the levels of the sun's atmosphere where selective radiation pressure may form the chief support of certain gases, the Doppler effect can enable upward-moving atoms of these gases to "climb out" of the absorption lines associated with them, and to be accelerated from the sun. Beyond the distance $r = (4/3)a$ approximately, the variation of velocity is given by

$$v^2 = v_\infty^2\left(1 - \frac{8}{7}\frac{a}{r}\right),$$

where v_∞ denotes the radial velocity attained by the particle at infinite distance. He found that for such atoms as those of singly ionized calcium, v_∞ is of order 1.6×10^8 cm/sec, roughly the same for all absorbing atoms, whatever the Fraunhofer line associated with them.

In this case the radial motion can readily be determined in terms of hyperbolic functions, but the formulae need not be given here. From them it appears that already at $r = 10a$ the radial speed attains 94% of its final value, so that over much the greater part of the journey to distance R the radial speed is almost uniform, as if there were no radial force (case *a*). At $r = 4a/3$, where the radial speed takes the atom out of its absorption line, that speed has about a third of its final value. This distance is attained in about 15 minutes. The distance R is attained in about 1.11 day, as compared with 1.09 day if the radial speed had the constant value v_∞ given by Milne. In this as in many other (but not all) respects, motion according to Milne's theory can be illustrated very approximately by the case (a), of uniform radial speed.

In Milne's theory the atom is continuously accelerated upwards, with ever diminishing force, against the pull of solar gravity. Soon it attains the escape speed, and then exceeds this speed. At the sun's surface the escape speed is $(2ga)^{1/2}$, where g denotes the surface value of solar gravity (2.74×10^4 cm/sec^2). Thus it is 620 km/sec; at any greater distance r it is less in the ratio $(a/r)^{1/2}$, so that at $10a$ it is 196 km/sec and at the earth's distance, 42 km/sec (this is much greater than the escape speed from the earth, so that most particles that escape from the outer layers of the earth's atmosphere are still held in the solar system).

Even if the accelerating force is negligible beyond $10a$ (and we are ignorant on this point), the proportionate change of speed beyond this distance, to infinity, due to the backward pull of gravity, is small if the speed at $10a$ is 500 km/sec or more. For example, for a speed there of 500 km/sec it is 8%, and for a speed there of 1000 km/sec it is only 2%. The mean speed of a particle that is accelerated uniformly, from rest at the sun's surface to a speed of 1000 km/sec at $10a$, and then travels freely against gravity, is about 960 km/sec.

2.2. Continuous emission from P; the "stream"-curve

If particles are continuously emitted from P, as above, those that have issued prior to any time t will at this instant lie on a certain curve. Each point on this curve will be occupied by a particle that issued at an earlier instant t_0, and t_0 is a parameter in terms of which (and of t) the coordinates r, θ of the point can be expressed. The polar equation of the curve can be obtained by eliminating t_0 from the equations for r and θ, that is, by substituting in (4) the value of t_0 given by (6). Thus the polar equation is:

$$\theta = \Theta_0 + \lambda + \Omega t + \omega_0 a^2 \int_a^r dr/r^2 f(r) - \Omega \int_a^r dr/f(r)$$

$$= \Theta_0 + \lambda + \Omega t + F(r), \tag{2.10}$$

where $F(r)$ is independent of λ and t; clearly $F(a) = 0$.

Like the path of any individual particle, the "stream"-curve, along which lie the particles continuously emitted from P, has a *form* which is independent of λ and T; their values merely determine the particular orientation of the curve in the sun's equatorial plane. The stream-curve revolves with the angular velocity Ω of the emitting point P.

The inclination χ of the stream-curve, to the radius through the point r,θ on it, is given by

$$\tan \chi = r d\theta/dr = r dF(r)/dr = - \{\Omega r/f(r)\} + \omega_0 a^2/r f(r).$$

At $r = R$, writing $\chi = \chi'$, we have

$$\tan \chi' = -(\Omega R/v)(1 - \omega_0 a^2/\Omega R^2). \tag{2.11}$$

The angle subtended at the sun's center O by the part of the stream-curve up to the distance R will be denoted by $\Delta\Theta$; it is the longitude of the emitting point P relative to the sun's central meridian (as seen from the earth), when the curve passes through this meridian. Clearly

$$\Delta\Theta = F(R) = \omega_0 a^2 \int_a^R dr/r^2 f(r) - \Omega \int_a^R dr/f(r). \tag{2.12}$$

The ecliptic, the plane of the earth's orbit round the sun, is only slightly inclined (at the angle 7°.3) to the equatorial plane of the sun; hence the azimuth θ' of the sun's central meridian (as seen from the earth) is given with close approximation by

$$\theta' = \Omega' t + \theta'_0, \tag{2.13}$$

where θ'_0 denotes the azimuth at zero time. The time t_c at which the emitting point P is on the central meridian, that is, at which $\Theta = \theta'$, is given by

$$(\Omega - \Omega')t_c = \theta'_0 - \Theta - \lambda. \tag{2.14}$$

If t' is the time at which the stream-curve passes through the earth's center, that is, when θ in (10), for $r = R$, is equal to θ' in (13), then

$$(\Omega - \Omega')(t' - t_c) = - F(R). \tag{2.15}$$

If the radial speed is constant, we have the following special forms of some of the above equations:

$$F(r) = -(\Omega r/v)(1 - \omega_0 a/\Omega r)(1 - a/r)$$

$$\Delta\Theta = -(\Omega R/v)(1 - \omega_0 a/\Omega R), \tag{2.12a}$$

$$t' - t_c = (1 - \omega_0 a/\Omega R)(\Delta t)/(1 - \Omega'/\Omega), \tag{2.15a}$$

the last two being approximate.

2.3. Numerical illustrations, for constant radial speed

The above formulae are here used to infer the values of various times and angles associated with the emission of particles from P, for the special case (*a*) in which no radial (or other) force acts on the particles after emission, so that the radial speed and the angular momentum of each particle are constant.

The radial speed is doubtless different from one emitting center to another, and for any one center it may vary with time. Hence several values of v are considered, from 100 to 2000 km/sec, or 10^7 to 2×10^8 cm/sec (Figs. 1, 2).

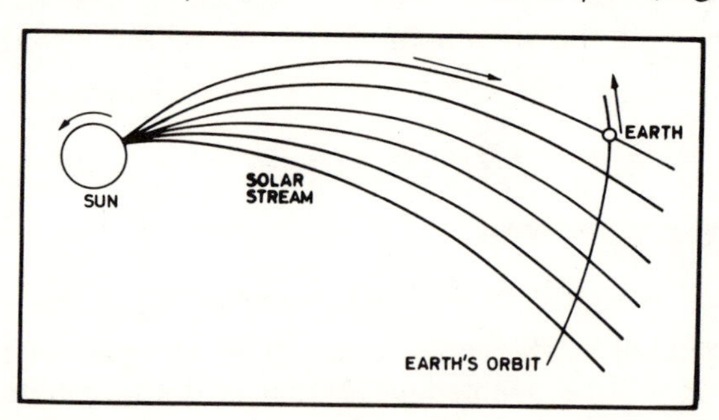

Fig. 2.1 Sketch illustrating a solar stream of particles moving with a speed corresponding to a time of travel of 36 hours from the sun to the earth. (Chapman and Bartels, 1940, p. 413.)

Table 1 contains particulars relating to the paths of individual particles; v is given in row 1, and the corresponding travel times Δt to the earth (in days) in row 2; they range from 17 days to 21 hours. Rows 3, 4, 5 refer to emission with lateral speed equal to the speed of rotation of the sun's surface at the equator, 2 km/sec, corresponding to $\omega_0 = \Omega$; rows 6, 7, 8 refer to emission with lateral speed $\pm$ 100 km/sec, corresponding to $\omega_0 = \pm 50\Omega$; here the plus sign corresponds to lateral motion in the same direction as that of the solar surface, and the minus sign to the opposite direction. Values are given of the inclination ϕ of the path to the radius, at emission (ϕ_0, for $r = a$) and at the earth's distance (ϕ', for $r = R$). The values in rows 3 and 4, for emission with no lateral speed relative to P, show

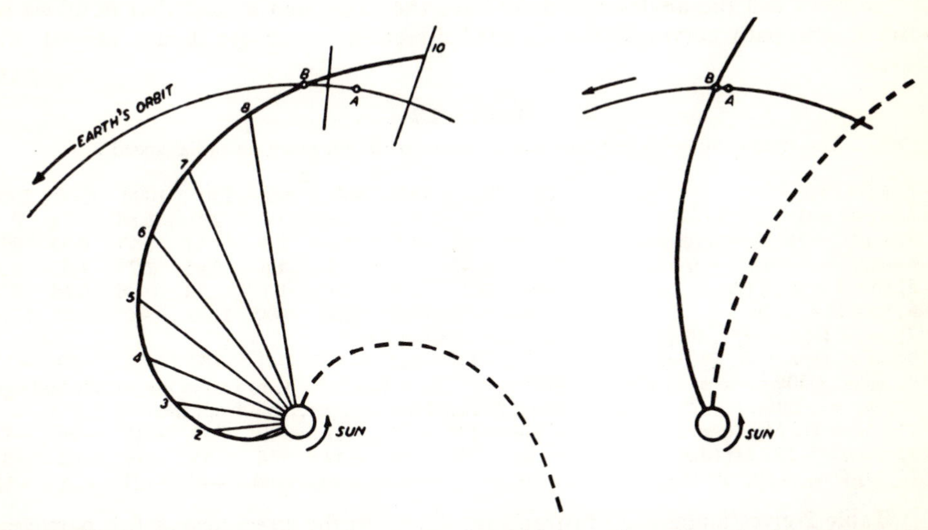

Fig. 2.2 Illustrating the curvature of a stream of particles, continuously emitted radially from a point on the sun, owing to the sun's rotation. The radii of the sun and of the earth (indicated by the small circles A and B, at two points along its orbit) are shown magnified tenfold, in relation to the distance (the orbital radius) between them. On the left the speed of the particles is taken to be 200 km/sec, and on the right, 1000 km/sec. On the left the continuous heavy line represents the particles, and the numbers along the line indicate the positions of particular particles emitted at equal successive intervals of time; each such particle, since its emission, has travelled along the radial line shown. (after J. Bartels).
(Geomagnetism, p. 384.)

that in this case the path is almost radial at all distances. The values in row 6, for lateral speed 100 km/sec, show that for the smaller values of v the direction of emission is considerably oblique to the radius; but those in row 7 show that at the earth's distance the path is almost radial, for all the radial speeds considered.

TABLE 2.1

The path of a particle with constant radial speed v

Row 1; v (km/sec)	100	200	300	400	500	750	1000	1500	2000
Row 2; travel time (days) Δt	17.4	8.7	5.8	4.3	3.5	2.3	1.74	1.16	0.87*
Row 3; $\omega_0/\Omega = 1$; ϕ_0 (deg)	1.14	0.57	0.38	0.28	0.23	0.15	0.11	0.08	0.06
Row 4; $\omega_0/\Omega = 1$; ϕ_1 (deg)	In all cases less than 0.01°								
Row 5; $\omega_0/\Omega = 1$; $\Delta\theta$ (deg)	In this case $\Delta\theta = \phi_0$ (see row 3) within 0.01°								
Row 6; $\omega_0/\Omega = \pm 50$; ϕ_0 (deg) $= \pm$	45.0	26.6	18.4	14.2	11.3	8.2	5.7	3.8	2.9
Row 7; $\omega_0/\Omega = \pm 50$; ϕ' (deg) $= \pm$	0.27	0.13	0.09	0.07	0.05	0.04	0.03	0.02	0.01
Row 8; $\omega_0/\Omega = \pm 50$; $\Delta\theta$ (deg) $= \pm$	57.3	28.6	19.1	14.3	11.5	7.6	5.7	3.8	2.9

* 20.6 hours.

Table 1 also gives the angle $\Delta\theta$ subtended at the sun's center O by the part of the path up to the earth's distance; for zero relative lateral speed of emission, this angle is almost negligible (row 5); for the lateral speed of emission 100 km/sec, the angle subtended is considerable for the lower radial speeds (row 8).

In rows 6–8 the angles ϕ and $\Delta\theta$ have the same sign as ω_0; that is, if ω_0 is positive, the path curves in the forward direction; otherwise it lags behind the radius.

TABLE 2.2

The stream-curve for particles uniformly emitted; for constant radial speed v

Row 1; v (km/sec)	100	200	300	400	500	750	1000	1500	2000
Row 2; $\omega_0/\Omega = 1$; $t' - t_c$ (days)	18.6	9.3	6.2	4.65	3.7	2.5	1.86	1.24	0.93
Row 3; $\omega_0/\Omega = 50$; $t' - t_c$ (days)	14.3	7.15	4.7	3.6	2.9	1.91	1.43	0.95	0.71
Row 4; $\omega_0/\Omega = -50$; $t' - t_c$ (days)	22.9	11.45	7.6	5.7	4.6	3.05	2.29	1.53	1.14
Row 5; row 4–row 3; (days)	8.6	4.3	2.9	2.1	1.7	1.14	0.86	0.58	0.43
Row 6; row 6 ÷ 32 sec	23,000	11,500	7700	5700	4600	3100	2300	1500	1150
Row 7; $\omega_0 = \Omega$; $r = a$; χ_0 (deg)	Zero for all speeds v								
Row 8; $\omega_0 = \Omega$; $r = R$; χ' (deg)	−77	−65	−55	−47	−41	−30	−23	−19	−12
Row 9; $\omega_0 = \pm 50\Omega$; $r = a$; χ_0 (deg)	Within 2%, $\chi_0 = \phi_0$ (Table 1, row 6) in magnitude and sign								
Row 10; $\omega_0 = \pm 50\Omega$; $r = R$; χ' (deg)	These angles differ inappreciably from those in row 8								
Row 11; $\omega_0 = \Omega$; $\Delta\Theta$: (deg)	−246	−123	−82	−61	−49	−33	−25	−16	−12
Row 12; $\omega_0/\Omega = 50$; $\Delta\Theta$ (deg)	−189	−95	−63	−47	−38	−25	−19	−12	−9
Row 13; $\omega_0/\Omega = -50$; $\Delta\Theta$ (deg)	−302	−151	−101	−75	−60	−41	−31	−20	−15

Table 2 gives numerical illustrations relating to the *stream-curve* for particles continuously and uniformly emitted from a solar point P, and moving thereafter free from any force. Row 1 gives the same constant radial speeds as in Table 1. Rows 2–4 give values, for different radial and lateral speeds of emission, of the time interval between the impact of the particles on the earth, and the passage of P across the central meridian of the sun, as seen from the earth. Row 2 corresponds to zero lateral speed of emission relative to the motion of P, namely to $\omega_0 = \Omega$; in this case the first factor in (15a) differs inappreciably from 1, and $t' - t_c$ consequently differs from Δt (Table 1, row 2) only by the last factor in (15a); this gives an increase of 7%. When, however, the lateral speed is considerable, as when $\omega_0 a$ is ± 100 km/sec, $t' - t_c$ is materially affected by the first factor in (15a), being decreased if the lateral speed is forward (row 3), and increased if backward (row 4).

Rows 7–10 relate to the inclination of the stream-curve to the radial direction; initially, for $\omega_0 = \Omega$, the emission is radial ($\chi_0 = 0$, row 7), but at the earth's distance (row 8) the inclination, especially for the smaller radial speeds, is considerable; with increasing distance it tends to $-90°$, where the minus sign signifies that the stream-curve lags behind the radial direction. For lateral speeds of 100 km/sec, to which rows 9, 10 refer, the initial inclination χ_0 is almost equal to ϕ_0 (Table 1, row 6), and has the same sign; that is, the stream-curve bends forward initially if the initial lateral speed is forward; with increasing distance the forward inclination decreases to zero, at $r/a = (\omega_0/\Omega)^{1/2}$, or, for lateral speed 100 km/sec, at 7.1 solar radii; beyond this distance the stream curve bends backward, and finally lags behind the radius, though less than if the initial lateral speed were backward. At the earth's distance the inclination is practically the same as for relative radial emission (row 10 is thus the same as row 8).

Rows 11–13 give the angle subtended by the stream-curve, up to the earth's distance, at the sun's center O; it is large for the smaller radial speeds, and depends appreciably on the lateral speed at emission.

2.4. Emission within a conical solid angle

So far the emission from the point source P has been considered as in a single direction, which depends on the ratio of the lateral to the radial speed at the sun's surface. It is likely, however, that the emission is in all directions within a certain conical solid angle. In this case the stream will lie within the surface formed by the stream-curves for the limiting directions. In the equatorial plane the stream will have an angular breadth at the earth's distance proportional to the time difference between the limiting times of passage across the earth's center, $t'_2 - t'_1$, where the subscripts 1, 2 refer to the forward and rear stream-curves. This is illustrated by rows 5, 6 of Table 2, for the case when the limiting curves correspond to lateral speeds ± 100 km/sec, with radial speed v. Row 6, giving the differences between the values of t' in rows 4 and 5, gives the corresponding difference between the times of passage across the earth—or, what is equivalent, the duration of the interval in which the earth is immersed in the stream; this interval is 8.6 days for the slowest radial emission (100 km/sec) and for the fastest in Table 2, 2000 km/sec, the interval is less than half a day. As the stream sweeps across the earth's diameter in 32 seconds, by dividing the values in row 5 (there given in days) by 32 seconds, we obtain the breadth of the stream, at the earth's distance, in earth-diameters (Table 2, row 6).

2.5. Transported solar magnetic fields and cosmic rays

The above very simple calculations on the geometry and kinematics of emission of gas from the sun gain increased interest in view of present-day conceptions of the dragging-out of lines of magnetic force, by the (ionized) gas, from near the sun's surface into interplanetary space. Such lines of force may lie nearly along the stream, and may scatter, guide or constrain solar cosmic rays. If so, these rays may approach the earth, when this is immersed in such a stream, along the direction of the stream-curves; this direction may differ appreciably from the radial direction, that is, from the direction towards the sun; the difference of direction is measured by the angle χ', illustrated in Table 2 (rows 8, 10).

2.6. Attempts to find the radial speed by the superposed epoch method

If the emitting sources P could be identified, we could observe their times t_c of passage across the central meridian of the sun as viewed from the earth. If geomagnetic disturbance were observed following such central passage, one might estimate the time t' at which the forward limit of the stream crosses the earth. Then, if the emission is on the average radial, and the emitting area and its conical angle of emission are small, the travel time Δt of the gas for passage from sun to earth may be estimated, and hence also the average radial speed v. Clearly the inference involves several uncertainties. However, as yet we cannot identify the stream sources on the sun, so that t_c cannot be observed.

To meet the difficulty, recourse has been had to the method of superposed epochs, using the daily sunspot numbers R and the daily magnetic character figures C. Days of notable activity C are taken as the selected epochs, and the sunspot numbers are tabulated for these days ($n = 0$), and for several days before and after. Averages of the columns are then taken. Maurain (1934) thus found that the graph of the average R has a rather flat maximum about $2\frac{1}{2}$ days before the selected epochs. This interval is taken to represent $t' - t_c$, supposing that the emitting sources lie among the sunspots. For radial emission this implies a mean radial speed of about 750 km/sec. In this investigation the average level of magnetic activity at the selected epochs was only moderate, so that the inferred speed may refer to rather weak and relatively slow streams. Stagg (1929) applied the superposed epoch method to stronger magnetic activity, and obtained intervals of respectively 1.1 and 1.5 days. These correspond to speeds of order 1500 km/sec (Fig. 3).

2.7. The lateral expansion of a stream in transit from a point source

Subject to the validity of the assumption that the particles are free from force after emission, the outer surface of a solar stream emitted from a point source P

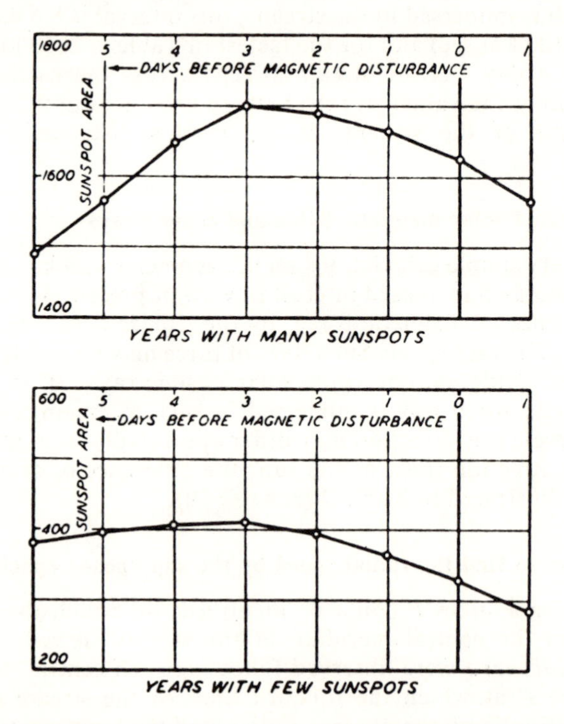

Fig. 2.3 Average sunspot areas, expressed in millionths of the sun's disk, from 6 days before until 1 day after days (epoch 0) selected as magnetically disturbed. Above: averages for 250 sets of days in years with many sunspots. Below: averages for 116 sets of days in years with few sunspots. (after J. M. Stagg, 1928.)
(Chapman and Bartels, 1940, p. 384.)

on the rotating sun is determined by the conical solid angle of emission. Practically nothing is known with certainty about this, as we do not understand the cause and mode of emission. It may be that the conical angle is hemispherical, in the sense that the emission is in all directions above the tangent plane to the solar surface at P; this is so, for example, for the gas escaping into a vacuum through a small hole in a plane boundary of a vessel containing gas under pressure (it is not suggested that this is a good model of the process of emission from the sun). The velocity distribution at emission must, in fact, be represented by a distribution of points in a velocity space extending over the whole space on one side of a plane parallel to the tangent plane at P; the previous discussion, considering a single direction and speed of emission, refers to only one of such velocity points. It is merely a simplification to speak of a conical angle of emission, and of a surface of the stream; but the simplification may well be reasonably appropriate to actual streams. Adopting it, we may see from the numerical illustrations given in Tables 1 and 2 how the stream broadens out, owing to purely geometrical and kinematic causes, from an initial conical angle of emission from a point source P. Thus, for example, considering (as in those Tables) only the spread in the sun's equatorial plane, emission between ϕ_{01} in the forward direction and $-\phi_{02}$ in the backward direction corresponds to an angle of emission $\phi_{01} + \phi_{02}$. The angles ϕ_{01} and ϕ_{02} are related to the corresponding values of ω_0, namely ω_{01} and $-\omega_{02}$, by

$$\tan \phi_{01} = \omega_{01}a/v, \quad \tan \phi_{02} = \omega_{02}a/v.$$

The angle subtended by the stream at the earth's distance is the difference between the corresponding values of $\Delta\Theta$. For constant radial speed this difference is obtainable from (12a), taking $r = R$ and using the two values of ω_0; it is $(\omega_{01} + \omega_{02})a/v$. If the two values of ω_0 are equal, we have

$$\text{expansion ratio} = \frac{\text{angle at } r = R}{\text{angle at emission}} = x/\arctan x,$$

where $x = \omega_0 av$. This is always greater than 1, but tends to 1 as x tends to zero (for example, as v increases). If we take $\omega_0 = 50$, as in Tables 1 and 2, the ratio is $4/\pi$ or 1.28 for v = a100 km/sec, and only 0.2% greater than 1 for $v = 2000$, The ratio in these cases is that of the difference between the numbers in rows 8 and 7 of Table 2, to those in row 6 of Table 1. As the former are given only to 1°, the ratio cannot be got very accurately from these Tables.

The lateral speeds of emission u from a point source may be partly thermal random speeds, as in the case of emission from a hole in the boundary of a vessel containing gas. In some early studies of magnetic storm theory the gas was thought of as being emitted with a temperature of about 6000°K, similar to that of the solar photosphere. As the gas is likely to be composed mostly of atomic hydrogen—the major constituent of the sun—this would correspond to a root-mean-square lateral speed, in any particular direction, of 0.091 $T^{1/2}$ km/sec. For $T = 6000°$ this is 7.05 km/sec. This speed u, acting to expand the stream outward in both the forward and backward lateral directions, would broaden the stream, in the course of the travel time Δt (Table 1, row 2), by the linear amount $2u\Delta t$; this is $1.22 \times 10^5 \Delta t$ km, if Δt is reckoned, as in Table 1, in days. For example, for radial speed $v = 100$ km/sec, for which Δt is 17.4 days, the thermal broadening of the stream would be 2.12 million km, or about 1660 earth diameters; for the lateral speed of 100 km/sec, as considered in Tables 1 and 2, the thermal broadening would be correspondingly greater, namely by about 23,000 earth diameters, as indicated in row 6 of Table 2. Such a lateral speed, 100 km/sec, would correspond to $T = 1.21 \times 10^6$; such temperatures are found in the corona. Kantrowitz (1960) and others have urged that the lateral speed is likely to be of much the same order as the radial speed, in any natural process of emission. For the greater radial speeds illustrated in Tables 1 and 2, this would correspond to temperatures of order 10^8; even such high temperatures are now considered as possibilities in solar and stellar phenomena.

2.8. The lateral expansion of a stream in transit from an emitting area

So far in this discussion the emission considered has been from a single point P. It must, of course, be from a finite area on the sun. Still dealing only with emission from the plane of the sun's equator, suppose that the area of emission on the sun extends in this plane over a range of longitude $d\lambda$. Equations (4, 4a) show that the range of the aximuth θ at any distance r from the sun is the sum of $d\lambda$ and of the breadth of the spread of the stream, corresponding to the last term in these equations; this corresponds to the broadening that results from the angular spread of emission from each point P. Thus at the earth's distance the breadth of the stream considered in, for example, row 6 of Table 2, must be increased by $Rd\lambda$, if $d\lambda$ is expressed in circular measure. The addition, reckoned in earth diameters, for values of $d\lambda$ equal to 0.1, 0.2, 0.3 radian (or 5°.73, 11°.46, 17°.19), has the values 1180, 2360, 3540. For the *greater* radial speeds considered in Tables 1,2, these additions are comparable with or greater than the breadths given in row 6 of Table 2. For the very slow streams they are less than the breadths there given. The longitude range of the emitting area will correspondingly increase the duration of the immersion of the earth in the stream, beyond such times as are given in row 5 of Table 2. For example, a stream with the range of lateral speed between ± 100 km/sec, corresponding to row 5 of Table 2, and with radial speed 1000 km/sec, could enclose the earth for $0{\cdot}86 + 0.076\, d\lambda^\circ$ days, where $d\lambda^\circ$ denotes the value of $d\lambda$ in degrees. Thus for a stream emitted from an area whose diameter (in solar longitude) is 10°, the duration, for the said range of lateral speed, would be 1.62 days.

If the emitting area is circular in form (though we are ignorant on this point), it will have a range l in solar latitude equal to its range in longitude. On this account the extent of the stream at the earth's distance, *perpendicular* to the equatorial plane, will be $Rdl = Rd\lambda$, values of which have been already cited). For emission with its center line radial, from all points of the area, there will be a conical expansion in the ratio $(R/a)^2$ or 45,400, and a corresponding decrease in density. There will be further expansion and decrease of density owing to the spread of directions of emission, from each point. Thus we may expect that the stream density may decrease from the sun's surface by a factor of order 10^5.

2.9. Emission from non-equatorial areas on the sun

Sunspots appear on the sun in solar latitudes up to or even beyond 30°, though most often between 10° and 15°. Sunspots are not an essential accompaniment of emission, because magnetic disturbance sometimes occurs when there is no visible spot on the sun. But there is undoubtedly a considerable association between magnetic disturbance and the presence of sunspots near the center of the solar disc, as seen from the earth. If uniform emission occurs from a solar area not on the equator, the center-line of the stream will describe a non-planar surface of revolution, unless the central direction of emission is parallel to the equatorial plane. For radial speeds of 1000 km/sec or more, the stream will, however, be only moderately curved, up to the earth's distance; and its center-line will describe a nearly conical surface. The rate of revolution of the stream around the sun's axis is Ω, whatever the area or direction of emission, so long as the emission is uniform. The position and direction of the stream at the earth's distance depend mainly on the direction of emission, and only to a minor degree on the precise latitude of the emitting area. Until there is observational or theoretical evidence to the contrary, it is natural to assume that on the average the emission is radial, though in particular cases there may be considerable deviations from this.

The ecliptic plane (of the earth's orbit) is inclined at 7.3° to the plane of the sun's equator. The earth crosses this equatorial plane on about June 5 and December 5, that is, near the solstices. Its heliographic latitude l' has its maximum values, 7.3° north and south, respectively on about September 5 and March 5. Clearly, near September 5 streams radially emitted from regions in northern solar latitudes have a better chance of sweeping over the earth, than streams from southern latitudes; a much greater conical angle of emission and area is required in the latter case, and conversely near March 5.

The frequency and intensity of geomagnetic disturbance have two annual maxima, near the equinoxes, with intervening minima near the solstices. These maxima are simultaneous in the northern and southern hemispheres of the earth. Hence they are not due, as the terrestrial seasons are, to the annual change in the orientation of the earth relative to the sun—or, what is equivalent, to the changing geocentric direction of the sun. It seems more likely that they are connected with the changing orientation of the sun relative to the earth, or, what is equivalent, to the changing heliocentric direction of the earth. Tandon (1956) has given tables of *sequences* of magnetic disturbances which he identifies with areas on the sun where long-continuing but intermittent sunspot activity has appeared; some of these areas lie in the northern, others in the southern hemisphere of the sun. His tables lend support to the idea that northern spot-areas are most effective when the earth's heliocentric direction is north of the sun's equator, and southern areas when the earth is south of the sun's equator.

The theoretical formulation of the statistics of recurring geomagnetic storms is inevitably rather complex. It may be attempted on a simplified basis as follows.

Let $f(T, l)$ be the distribution function for the occurrence at sunspot epoch T, in heliographic latitude l, of continuously emitting sources of solar streams. The inclusion of the variable T implies an assumption that the emitting areas vary in latitude distribution in the course of the sunspot cycle, as the sunspots themselves are known to do. The mean latitude of sunspots decreases from one sunspot minimum to the next; at each minimum there is a discontinuity of sunspot latitudes, and for a time there may be higher latitude spots of the new cycle coexisting with low latitude spots of the old cycle.

Let $F(T, l, \alpha)$ denote the distribution function of the conical angles of emission that are greater than or equal to 2α, for sources at latitude l at sunspot epoch T. Let l' denote the heliographic latitude of the earth; this is a function of date in the calendar year.

The frequency of sequences of recurring geomagnetic disturbances near June 5 and December 5, when l' is small or zero, will be proportional to

$$\int_{-90^\circ}^{90^\circ} f(T, l) F(T, l, l)\, dl, \tag{2.16}$$

because a stream with conical angle at least equal to 2α, from latitude l, can sweep the earth at that time only if $\alpha = l$.

The corresponding frequency at or near September 5, when l' is $7^\circ.3$ north, is

$$\int_{0}^{90^\circ} f(T, l) F(T, l, l - 7.3) dl + \int_{-90^\circ}^{0} f(T, l) F(T, l, l + 7.3) dl. \tag{2.17}$$

The ratio of these two quantities may be taken as a measure of the annual frequency variation of recurrent geomagnetic disturbance. This ratio may here be illustrated in a rather grossly simplified way. We suppose that at sunspot epoch T all emitting sources are in a definite latitude $l(T)$, and that the function F is independent of T and l, and corresponds to a Gaussian distribution of 2α with mean value $2\alpha_0$. Then F is proportional to

$$\int_{x}^{\infty} e^{-(x/x_0)2} dx, \quad \text{where} \quad x_0 = (\pi)^{1/2}\alpha_0. \tag{2.18}$$

Thus the frequency ratio solstice/equinox is

$$\frac{2F(l)}{F(l - 7.3) + F(l + 7.3)} \tag{2.19}$$

where l denotes $l(T)$. Table 3 illustrates the values of this ratio for some chosen values of α_0.

TABLE 2.3

$$A = \frac{2F(l)}{F(l - 7^\circ.3) + F(l + 7^\circ.3)}; \quad l = 10^\circ$$

$\alpha_0{}^\circ$	5	10	15	20	25	30	40	50
A	0,332	0,88	0,955	0,982	0,99	0,99	0,99	0,99

From Table 3 one can infer that for l equal to 10°, if the streams have a mean conical angle $2\alpha_0$ of emission amounting to 30° or more, the semi-annual variation of disturbance frequency would only be a few per cent; but as $2\alpha_0$ decreases below 30°, the expected semi-annual variation of disturbance would rapidly increase.

Priester and Cattani (1962) have made more detailed analysis of this question, and their work supports the idea that the semi-annual variation of magnetic activity is caused by the changing heliocentric latitude of the earth relative to the main zones of sunspot appearance.

2.10. The nature of the onset of a stream-produced magnetic storm

The density of particles in a solar stream of the kind here discussed is likely to fall off towards the lateral boundary of the stream. Hence as the stream overtakes the earth, the gas in which the earth becomes immersed will at first be less dense than when the central part of the stream pours upon the earth. For this reason we may expect that magnetic storms produced by such streams will begin gradually.

This is consistent with the observation that the 27-day recurrence tendency which stream-produced storms may be expected to exhibit is indeed manifested by gradually commencing storms, but not to anything like the same degree by storms with sudden commencements. The greater the intensity of storms, the more commonly do they have a sudden commencement, and show no 27-day recurrence (Greaves and Newton 1929). Hence we may expect that storms of this latter type are caused by a solar agent differing from the solar streams thus far discussed. They are, in fact, found to show a considerable correlation with solar flares.

Solar flares are sudden brightenings of a small part of the sun's surface. The area of enhanced brightness is often within the region of a complex changing sunspot group. The brightening often progresses along a sinuous path in such a region. The rise to maximum brightness is usually more rapid than the decline; the whole flare may last from a few minutes to times of the order of one or two hours. Along with the visible light from the sun, that reveals to us the occurrence of the flare, there comes ultraviolet light. Simultaneously with our view of the flare this short-wave light increases the ionization of the atmosphere, chiefly in the lower E and the D layer. This affects radio communication in different ways, depending on the radio wave frequency. Most intense flares observed near the central part of the sun's disc, as seen from the earth, are followed by a suddenly commencing magnetic storm after an interval of about a day; sometimes this occurs even for flares situated close to the edge or "limb" of the sun. In the latter case the emission of the solar gas responsible for disturbing the earth must either be very obliquely directed relative to the radius through the flare, or, what seems more likely, the conical angle of flare emission of gas is large, not much less than that of a hemisphere. Following Kahn (1949), we next discuss the geometry and kinematics of the gas emitted from a flare.

2.11. The shell of solar gas emitted from a solar flare

A typical flare has an area of 10^{-3} of the sun's apparent disc; its mean radius b is therefore about equal to 2×10^9 cm. Its duration t_1 may be taken as 10^3

seconds, or about $\frac{1}{4}$ hour, though some last several times as long. Let the surface of the flare be regarded as a hemisphere of radius b, from which, during the time interval of length t_1, particles are emitted within a solid angle Ω at the rate $N(t)$ per second per unit area, reckoning t from the beginning of the flare. The speed of the emitted particles is taken to be a function $v(t)$ of time; the flare brightens more rapidly than it declines. For simplicity suppose that the maximum brightness is attained instantaneously, and that the impulsion given to the particles decreases as the flare fades. Then $v(t)$ decreases with t, from its initial value v_0 to its final value v_1; taking the decrease to be linear,

$$v(t) = \{(t_1 - t)v_0 + tv_1\}/t_1. \tag{2.20}$$

The speed v is taken to be constant after emission.

The gas travels outward as part of a spherical shell centered on the flare. It will be called a (flare-)*shell*. Because of the brief duration of a flare, the shell, unlike the *streams* previously discussed, is not significantly affected by the rotation of the sun.

The shell first reaches the earth's distance R at the time R/v_0 from the flare commencement, and passes beyond this distance at the time $R/v_1 + t_1$. Thus the earth is immersed in the shell, if this passes over the earth, for the time $R/v_1 - R/v_0 + t_1$. If, for example, v_0 is 1500 km/sec and v_1 is 1000 km/sec, the duration is 5.1×10^4 sec, or about $14\frac{1}{2}$ hours, and it begins 10^5 sec (27.7 hours) after flare commencement.

The particles emitted between the times t and $t + dt$ will at a later time t' occupy a shell of radius r given by $b + (t' - t)v$, and of thickness dr given by

$$dr = \{(t' + t_1 - 2t)v_0 - (t' - 2t)v_1\}(dt)/t_1. \tag{2.21}$$

The density of gas in this shell, $n(r, t')$, is $b^2N(t)\,dt/r^2\,dr$; in this expression t and dt are connected with r, t' (and b, v_0, v_1, t_1) by the relations already given. When $t' - t$ is small, that is, during the early history of the shell (and especially during the flare itself), $n(r, t')$ is approximately $b^2N(t)/r^2v$, proportional to the inverse square of the distance. For large values of t' the density is approximately $b^2N(t)t_1/r^2t'(v_0 - v_1)$; its values when the shell first reaches the distance r, and when it just passes beyond this distance, are respectively

$$b^2\nu_0 t_1 v_0^2/R^3(v_0 - v_1) \quad \text{and} \quad b^2\nu_1 t_1 v_1^2/R^3(v_0 - v_1), \tag{2.22}$$

where ν_0 and ν_1 denote the density of the emitted gas at the beginning and end of the flare. Thus the ratio of the initial to the final density of the shell at distance R is $\nu_0 v_0^2/\nu_1 v_1^2$; for the adopted illustrative numerical values, this is $2.25\ \nu_0/\nu_1$. The initial density at this distance is $5.3 \times 10^{-10}\nu_0$. For example, if ν_0 is 10^{11}, the shell will reach the earth's distance R with density 53 particles per cm^3; but we have no information as to the value of ν_0.

Though the flare does not attain its maximum brightness instantaneously, the particles of the shell that first reach the distance R may still correspond to the time and speed of maximum emission and brightness. If so, the initial density at this distance will be

$$\frac{b^2\nu_0 v_0^2}{R^3}\left\{\frac{t_1'}{v_0 - v_1'} + \frac{t_1''}{v_0 - v''_1}\right\}. \tag{2.23}$$

Here v_1' and v_1'' denote the initial and final values of v (the speed of emission), at intervals t_1' and t_1'' respectively before and after the epoch of maximum emission speed v; thus t_1 the whole duration of the flare is now $t_1' + t_1''$, and if $v_3' = v_1''$, the initial density at R is still given by the former expression.

Clearly the immersion of the earth in the solar gas is likely to be much more sudden for a plasma-shell from a flare than for a stream from an M region. The wider angle of emission from a flare than from a stream is inferred from the occurrence of magnetic storms following flares situated at almost any distance from the center of the sun's disc, as viewed from the earth.

Though flares generally occur in the midst of actively changing sunspot groups, flare-produced suddenly-commencing magnetic storms show little 27-day recurrence tendency. This is explained by the slightness of the relation between the time of occurrence of flares and the passage of the sunspot group across the central meridian of the sun.

2.12. Retardation on the way from the sun

The fast-moving gas that produces magnetic storms traverses interplanetary space, which is not empty. Hence it is relevant to consider whether the gas suffers much retardation on its way to the distance R of the earth. The interplanetary gas may be taken to be mainly ionized atomic hydrogen, that is, protons and electrons. The only perceptible retardation will be that caused by the interplanetary protons. The proton-proton collision cross-section at temperature T is of order $3 \times 10^{-5}/T^2$; for stream or cloud particles having a speed of 10^8 cm/sec or more, T will be effectively of order 10^8, so that the collision cross-section is of order 3×10^{-21}. The number-density of the interplanetary gas falls below $10^8/\text{cm}^3$ at a few solar radii, that is, at a very small fraction of the distance R. The law of variation of this number density n with distance r from the sun is not properly known, but for the purpose of this calculation may be taken to be as $1/r^3$. Hence $\int ndr$ beyond, say, two solar radii (2a) is of order $10^8 a$ or about $10^{19}/\text{cm}^2$. Thus the chance of a collision between $r = 2a$ and $r = R$ may be estimated as $10^{19} \times 3 \times 10^{-21}$ or about 1/30. Hence it appears that retardation in interplanetary space between the sun and earth will at most be moderate. Nearer the sun the particles of the stream or shell may suffer collisions in the denser gas there existing; but the process of ejection must suffice to make the particles leave the neighbourhood of the sun—limited by a few solar radii—with a speed of order 1000 km/sec or more (for the streams and shells that cause active magnetic disturbance). Weaker disturbance may be caused by slower rarer gas, more liable to retardation—because it will have a smaller effective temperature (calculated by the relation $\frac{1}{2}\text{mv}^2 = \frac{1}{2}kT$, where m denotes the mass of a hydrogen atom or proton). Hence its collision cross-section will be greater.

The general background gas in the interplanetary space—the extension of the solar corona, the irregular outer atmosphere of the sun—seems likely to have an outward motion. Biermann (1951, 1961) infers this from the study of comet tails. Parker (1961) and Chamberlain (1961) have made different estimates of the outward speed, on the basis of kinetic theory. This "solar wind", as Parker has named it, should lessen the retardation of the streams and clouds that produce magnetic disturbance.

The angular velocity of the interplanetary solar atmosphere is not known, but the flattening of the zodiacal light towards the plane of the ecliptic suggests that the gas and its associated dust are rotating (Chapman 1961). Lüst and Schlüter (1955) have investigated the influence of the sun's general magnetic field upon this rotation, and have inferred that the magnetic field may force the interplanetary gas—which is likely to be almost completely ionized—to rotate with the sun up to 50 solar radii. Sunspot magnetic fields are often much stronger than the general solar magnetic field, but they extend outwards less far; within their range they are likely to carry the solar atmosphere with them, except when local forces of other kinds are strong, such as impel upward prominence motion.

It seems desirable to use the three terms solar *stream*, solar *shell* and solar *wind* with discrimination, with definite meanings, namely, (*i*) a limited-angle emission from an M region, continuing for days or weeks, (*ii*) a wider-angle, radially limited emission from a flare, and (*iii*) a general outflow from the sun, at all times. All three cases come under the inclusive term solar plasma(-flow).

2.13 Plasma kinetic density and magnetic field energy

Let H_k denote the intensity of a magnetic field whose energy density is equal to the kinetic energy of a streaming neutral plasma, whose ion number density is n and whose ionic and electronic speeds are v. Then

$$H_k^2/8\pi = \tfrac{1}{2}nmv^2,$$

where m denotes the mass of an ion-electron pair. Taking the gas to be hydrogen, $m = 1.67 \times 10^{-24}$, so that if $v = 10^8$ cm/sec,

$$H_k^2 = 4\pi nmv^2, \quad H_k = 4.6 \times 10^{-4} n^{1/2}\,\text{gauss} = 46 n^{1/2}\gamma.$$

There is a current idea that when the magnetic field energy much exceeds the kinetic energy density in a plasma, the motion of the latter is dominated by the field and its changes, whereas when the kinetic energy density is much the greater, the plasma controls the field and can transport it. This conception can usefully be considered in relation to the solar plasma. The density of the plasma at the earth's distance, or at earlier stages in its motion from the sun. is not known. An estimate by Chapman and Ferraro (1932) is often quoted, though it was stated by them to be very tentative: it was that at the earth's distance n may range from 1 to 100, for plasma that causes moderate magnetic disturbance; higher values, perhaps exceeding 10^3, may correspond to very strong disturbance (and in this case v may be somewhat greater than 10^8 cm/sec). The corresponding values of H_k, in γ, are as follows:

	$v = 10^8$ cm/sec				$v = 3 \times 10^8$ cm/sec	
$n =$	1	10	100	1000	1000	10,000
$H_k(\gamma) =$	46	145	460	1450	4350	13,800

All these values of H_k exceed the actual field intensities normally measured by earth satellites beyond 10 earth radii, where (in the equatorial plane) the geomagnetic field intensity has fallen to 30γ.

Near the sun the density of the ejected plasma is likely to be greater by a factor of order 10^5 than near the earth, owing (as above shown) to conical expansion and thermal broadening. If v there has the values considered above, the corresponding values of H_k will be greater by a factor 3.15×10^2. This brings them into the range from 0.1 gauss upwards, and even, in the most extreme case given in the above Table, to 40 gauss. The latter is small compared with intensities of sunspot-magnetic fields in the photosphere and chromosphere, but it is large compared with the intensity of the general magnetic field of the sun (about 1 gauss near the surface at the poles).

The conclusions to be drawn from these comparisons are not clear, especially as regards the plasma in its earliest stages of ejection from the sun. This is partly because as yet we do not understand the process of ejection, in which strong local magnetic fields may play an important part.

2.14 The Electrical Discharge Theory of Solar Streams

Dr. C. E. R. Bruce, of the Electrical Research Association, at my request has kindly contributed the following account of his theory that the solar streams responsible for magnetic storms and auroras are produced by electrical discharges in the sun's atmosphere.

A theory of these solar and cosmic gas jets has been developed by Bruce (1957a, b, 1960, 1961, 1962b), and derives from his general thesis that most cosmic atmospheric phenomena have their origin in electrical discharges. One of the characteristics of these last is that the charged particles which carry the current tend to migrate inwards towards the axial regions under the influence of the discharge's own magnetic field. The gas pressure as a result is there increased by a factor which is proportional to the product of the current and the current density. He has emphasized that this increased pressure in the axial regions of a solar prominence has recently been observed (Bruce 1962a). When either the current or the current density varies, a pressure gradient, and hence a flow of gas along the "magnetic hose-pipe", so to describe it, formed by the discharge channel, will result. This he suggests is the cause of these cosmic gas jets. Their velocity is usually that of sound in ionized atomic hydrogen at the temperature of the gas, so that their temperature and velocities cannot be treated as independent, as so often has been done in the past. The resulting gas movements, as measured by the Doppler shifts of the bright emission lines emitted by the gas, therefore afford a "cosmic gas-velocity thermometer" (Bruce 1960a), the "readings" on which have been checked over the range 5000°K to 500,000,000°K. The upper limit of the range is set by the initiation of thermonuclear processes at about this temperature, which explains the corresponding upper limit of about 4000 km/sec thus set to cosmic gas velocities on which astrophysicists have commented (Bruce 1959, 1962b).

When these discharges are propagated radially outwards in the solar or stellar atmospheres, it appears that their temperatures can increase considerably, with a corresponding increase in the velocities of the jets they generate. For example, velocities of about 700 km/sec are observed just above the sun's surface, whereas the average velocity between sun and earth is about 2000 km/sec and particles are actually observed to enter the earth's atmosphere at 3500 km/sec. From the latter observation is deduced (Bruce 1959) that temperatures of over 100,000,000°K must be reached in solar outbursts, a deduction which was soon confirmed by satellite observations (Bruce 1960b).

It is suggested that a corresponding increase in the level of excitation of the gas as the discharge is propagated down a density gradient is observed in discharges in the very extensive atmospheres of the late type stars possessing combination spectra (Bruce 1956). A spectrum of relatively low levels of excitation (H, He I, and the singly ionized metals) is followed after a period of 100 to 200 days by one of much higher excitation reaching Fe X and Ca XV ". . . just as though they occurred as a consequence of the propagation of running waves over an extended medium." (Aller 1954.)

The theory would clear up the difficulty which Kantrowitz (1960) has emphasized is met by existing theories, for the velocity of the jet is intimately associated with its internal characteristics, but lateral spreading is prevented by the discharge's own magnetic field. The latter thus plays the important containing role which Chapman stressed would be a necessity in commenting on Kantrowitz's criticisms.

Some of the descriptions of the results of optical and radio investigations of the phenomena observed in disturbed regions of the sun's surface, are just those associated with electrical discharges. For example W. N. Christiansen and others (1960) write as follows: "The close association between the chromospheric plage, localized magnetic fields and the radio source suggests that the field plays a part in the production of these regions of high electron density. As suggested previously, it may be that the field provides a constraint, a sort of magnetic bag which prevents the material from dispersing." This will be seen to describe exactly the effect on an electrical discharge of its own magnetic field, for of course on the discharge theory the magnetic field and the high electron density are merely two aspects of the same phenomenon.

Indeed the electric discharge theory is the only one which explains the origin of the magnetic fields, the existences of which are so freely postulated by other theories of the origin of cosmic radio noise and of the structure of cosmic bodies—e.g. the spiral arms of the galaxies. The theory not only explains these last, but predicted the existence of similar structures on a stellar scale, a prediction which was soon verified by a reassessment of the existing photographic and spectroscopic evidence (Bruce 1961).

To sum up, solar jets are on this view not projectiles shot out from the sun, but are continuously generated from the gas of the sun's atmosphere all the way out from the sun by this magnetic pinch effect of the electrical discharge. As they are propagated outwards the increase in temperature observed near the sun's surface, from the 6000°K of the photosphere, through the 10,000°K to 20,000°K of the chromosphere, to the 1,000,000°K of the corona, is continued, until somewhere between the sun and the earth values of over 100,000,000°K are reached. The velocity of the jet is the velocity of sound in ionized atomic hydrogen at the discharge temperature.

It may be noted as further evidence in the theory's favour that it led to the conclusion that velocities of up to 7 or 8 km/sec should be observed in the Evershed effect in sunspots, despite the fact that for over 50 years since Evershed's own measurements, only velocities of 1 to 2 km/sec had been observed. However, the gas velocities predicted by the theory have recently been observed by Bumba in Russia.

3. The Mutual Influence of the Solar Gas and the Geomagnetic Field

3.1. Neutral ionized solar gas (or plasma)

As first indicated by Lindemann (1919), the solar gas that causes magnetic disturbances on the earth must be neutral but ionized. Atoms that are not ionized when they pass beyond the corona will be ionized by the sun's ultraviolet light on their way to the earth, and recombination will be insignificant. Sometimes thin wisps and arches of gas have been seen to rise through the sun's atmosphere with accelerating speed, sufficient to take them right away from the sun; these wisps and arches are seen by the light they emit, which indicates that their main constituent is atomic hydrogen. As they gradually become ionized they cease to be visible, because the scattering power of their electrons, such as visually indicates the presence of electrons in the corona, is insufficient to make them visible against their background.

The gas will contain a small proportion of helium, and some of the helium atoms may still be neutral. If so, they will travel onwards through the geomagnetic field into the earth's ionosphere, falling exclusively on the sunlit hemisphere. There they will add to the ionization, whose main cause is the sun's ultra-violet light. Solar eclipses offer a way of distinguishing between the parts of the ionization produced respectively by the wave radiation and by such neutral particles; a summary of work on this subject is given in Geomagnetism (Chapman & Bartels 1940). The beam of neutral particles will be eclipsed by the moon in such a way as to produce a solar "corpuscular" eclipse in the ionosphere, at a time and place considerably different from those of the "optical" eclipse. Such corpuscular eclipse decreases of ionization in the atmosphere have been repeatedly looked for, but with no clear result. The conclusion is that the neutral particle content of the solar streams and shells must be small. Hence in the following discussion the gas is supposed to be fully ionized. As it approaches the earth it must influence and be influenced by the geomagnetic field.

The following account summarizes in the first place the studies of this problem by Chapman and Ferraro (1931–1952). They dealt with solar gas free from any transported solar magnetic fields, and regarded the earth as devoid of any atmosphere in the regions penetrated by the solar gas.

The ionized gas is clearly a good electrical conductor. When a solid electric conductor approaches a magnet, electric currents are induced in the conductor by the magnetic field. These tend to shield the interior of the conductor from the field; also the interaction between the currents and field exerts a force that retards the advance of the conductor into the field. Such effects will develop also when a conducting *gas* approaches a magnet, but in addition the *form* of the gas will be modified by the electromagnetic interaction between the field and the currents.

For simplicity and definiteness let the front surface of the gas be supposed initially plane, perpendicular to the direction of motion of the gas. The gas is supposed to be of uniform density, and in the first instance all the particles in it will be supposed to have the same velocity, without superposed random thermal motions, or motion of divergence. Also let the velocity of the gas be parallel to the plane of the dipole equator. The earth's field is taken to be that of a point dipole at a point C, in a constant direction—although actually the dipole rotates daily around the geographic axis.

In this case the central point in the front surface of the gas, that lies on the line OC from the sun's center O to the dipole C, will experience the strongest geomagnetic field; the induced currents will be most intense there, and also the retarding force. Far from this point, in all directions along the surface, the geomagnetic field will be weaker, and likewise the induced currents and the retardation. Thus those parts of the surface will advance relatively to the central point, and a hollow will be formed round the earth. The earth's field is confined within this hollow space, called by Gold the *magnetosphere.*

3.2. The interaction between a thin plane conductor and the dipole field

Despite this inevitable distortion of the front surface of the gas it is instructive to consider the problem of the interaction between a thin plane conductor and the dipole field, when the conductor advances towards the dipole, with velocity w normal to the plane. Maxwell (1873), in his great treatise *Electricity and Magnetism*, discussed the equivalent problem of the currents induced in a thin plane conducting sheet ($z = 0$) by the advance towards it, with velocity $-w$, from the positive side ($z > 0$), of a magnetic distribution of any kind. The ratio of w to c, the speed of light, is supposed to be small, so that the displacement currents can be neglected. The equations of induction being linear, the effects of different magnetic systems are additive. Hence it suffices to consider the effect of a single magnetic pole, or of a dipole.

Let the magnetic pole be at the point P at a given instant, and let P' be the image point relative to the plane conductor: that is, the point along the normal through P to the plane, at the same distance on the negative side of $z = 0$. Let the specific electrical resistance of the sheet be denoted by $2\pi w_0$; w_0 has the dimensions of a velocity. Maxwell showed that the magnetic effect of the induced currents is to reduce the field of the pole at P, on the further side of the sheet ($z < 0$), in the ratio $w_0/(w + w_0)$; thus the field on that side is the same as if the sheet were absent and the strength of the pole at P were reduced in this ratio. On the side $z > 0$ the field of the pole at P is increased by the addition of the field of an (imaginary) image pole at P', of strength $w/(w + w_0)$ times that of P.

These results can clearly be generalized at once to the case of a dipole at P, in any direction: and to the case when the conducting sheet moves toward a stationary dipole, instead of the dipole moving towards the sheet.

When the sheet is perfectly conducting, so that w_0 is zero, there is perfect shielding from the magnetic field, on the side $z < 0$; and on the side $z > 0$, the field of the real dipole is increased by the field of an equal image dipole at P'.

It suffices here to suppose that the dipole at P is parallel to the sheet, in the $-y$ direction. Let the coordinates of P be 0, 0, c, and denote the magnetic moment by M. The dipole potential is $-My/r^3$, where

$$r^2 = x^2 + y^2 + z^2. \tag{3.1}$$

Maxwell showed that the current-function for the induced electric currents in the sheet is

$$-\frac{w}{w + w_0}\frac{My}{2\pi r_0^3},$$

where

$$r_0^2 = x^2 + y^2.$$

Thus the lines of current flow in the sheet are the intersections of the sheet with the equipotential surfaces of the dipole field. They are shown in Fig. 1, as they would be seen from the earth by an observer standing with the northward direction from his feet to his head, looking sunwards, towards a plane conducting sheet advancing to the earth normal to the sun-earth line. The horizontal line in the middle of the diagram lies in the plane of the dipole (magnetic) equator. Thus the direction of current flow there is towards the east. The lines of flow are the same, whatever the conductivity of the sheet; but this determines the *intensity* of the currents.

Fig. 2, like Fig. 1, illustrates the case when the sheet is perfectly conducting, so that the shielding of the dipole field on the rear side ($z < 0$) is complete. It shows roughly the form of the lines of force of the field in the plane containing the dipole, and normal to the sheet. In the geomagnetic case this includes the midnight and noon meridian planes. The combined field of the real and (equal) image dipoles has no component perpendicular to the sheet *at* the sheet. Thus the lines of force are tangential to the sheet, except at two singular points Q, where the lines meet the sheet perpendicularly; there the field intensity is zero. Thus the points Q are *neutral* points of the field. They are the focal points of the current system shown in Fig. 1, that is, the points around which the closed current flow circulates. The normal dipole field is, as it were, compressed by the conducting sheet, strengthening the field intensity in the equatorial plane, most of all near the sheet. In this plane the field of the image dipole doubles the dipole field intensity at the nearest point, on the sun-earth line; elsewhere on this line, the increase is less; its magnitude at the earth's center is $M/(2c)^3$, and at the nearer and further ends of the earth-diameter on this line it is $M/(2c - a)^3$ and $M/(2c + a)^3$, where a now denotes the radius of the earth.

Such an increase of horizontal magnetic force is observed at the beginning of a magnetic storm (Fig. 3). Its magnitude may be 20 γ or up to 100 γ in an intense storm. About half of this is due to currents induced in the earth by the sudden increase of the external field. Thus the amounts dH due to the latter, in the two cases, are 10 γ and 50 γ. Taking the earth's field intensity M/a^3, at the equator at the surface, to be 30000 γ, we can infer the distance c at which a perfectly conducting sheet, advancing towards the earth, would produce such an increase at the earth's center (or in the mean for the two ends of the diameter along the sun-earth line). Writing $c = fa$, we have

$$8f^3 = 30000/\,dH,$$

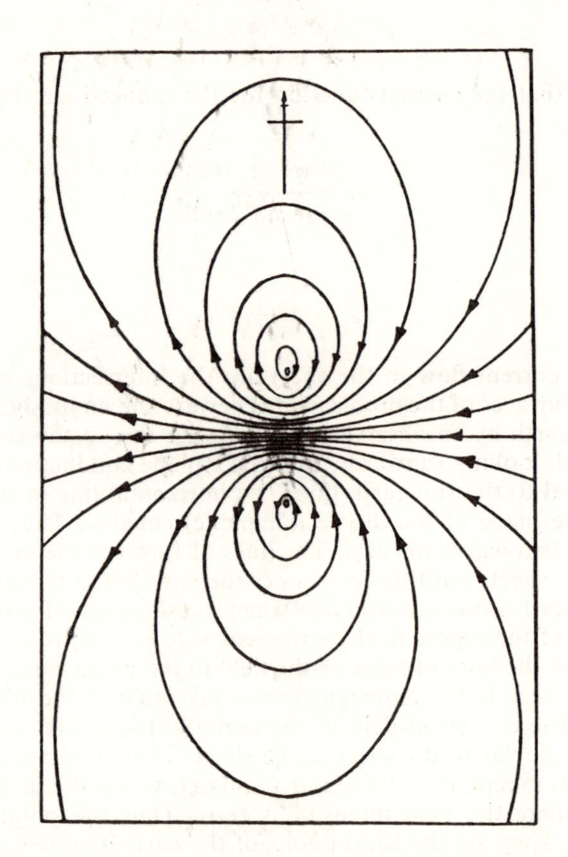

Fig. 3.1. Lines of flow of the electric currents induced in the plane surface layers of a neutral ionized stream advancing into the geomagnetic field, before the surface is materially distorted by the field. (Chapman and Ferraro, 1931, p. 177; also Geomagnetism, p. 860.)

if dH is expressed in γ. For $dH = 10$, this gives $f = 7.2$; for $dH = 50$, $f = 4.22$.

Thus the rapid approach of the solar gas towards the earth is seen to be likely to cause a rapid increase of the horizontal magnetic force, helping us to understand the mechanism of the first phase of a magnetic storm. Further, this very simple calculation indicates the order of magnitude of the distance of closest approach required to produce initial increases of the horizontal field, of the observed magnitudes.

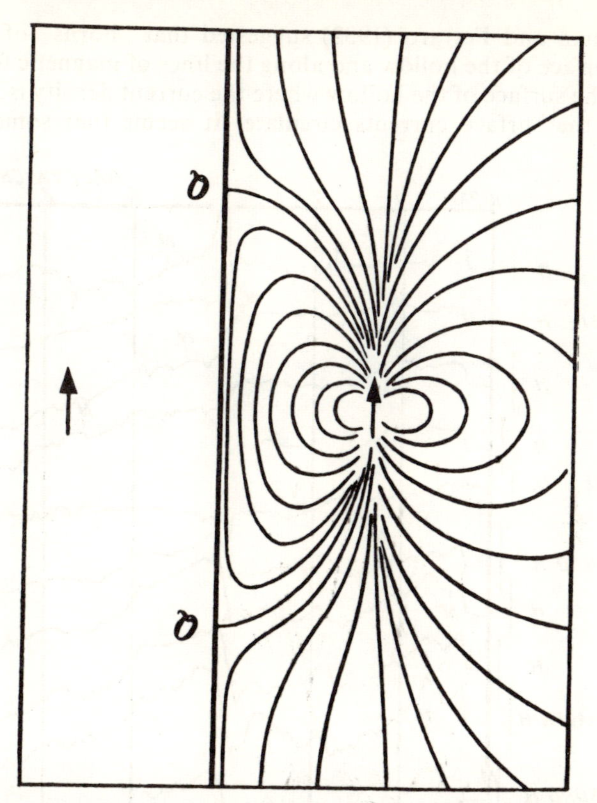

Fig. 3.2. Diagrammatic sketch of the combined field of the geomagnetic dipole and of the electric currents induced in the plane surface layers of the advancing neutral ionized stream, on the assumption that these currents completely shield the interior of the stream from the geomagnetic field; in this case the magnetic field of the induced current-system is the same as that of the image (shown on the right) of the geomagnetic dipole, relative to the plane face of the stream (Chapman and Ferraro, 1931, p. 179; also Geomagnetism, p. 861.)

The mechanical force caused by the action of the magnetic field H on the electric current flowing in the sheet conductor is $i \times H$ per unit area, where i denotes the surface-density of the current. The integral of this force over the sheet when this is at distance c from the dipole is $3M^2/(2c)^4$, the same as that between the dipole and an equal real dipole in the position of the image dipole. The force per unit area varies over the sheet, and only the postulated rigidity of the sheet enables it to remain plane. If the sheet, like the gas, were distortible, its surface would at once become curved, when it approaches the dipole field. The force vanishes altogether at the two neutral points Q; it is a maximum at the point of the sheet nearest to the dipole. Clearly the surface of the gas must bend round the earth, both in the plane of the dipole equator and in the meridian plane containing the dipole and the sun. But there would seem to be a tendency for the gas to flow onward through and near the points on the curved surface where this is met at right angles by lines of force of the field—the points around which the surface currents circulate. On this account Chapman and Ferraro (1932) suggested that "horns" of gas would flow out from the surface of the hollow and along the lines of magnetic force, around the two points on the surface of the hollow where the current density is zero—the points around which the surface currents circulate. It seems that some solar particles

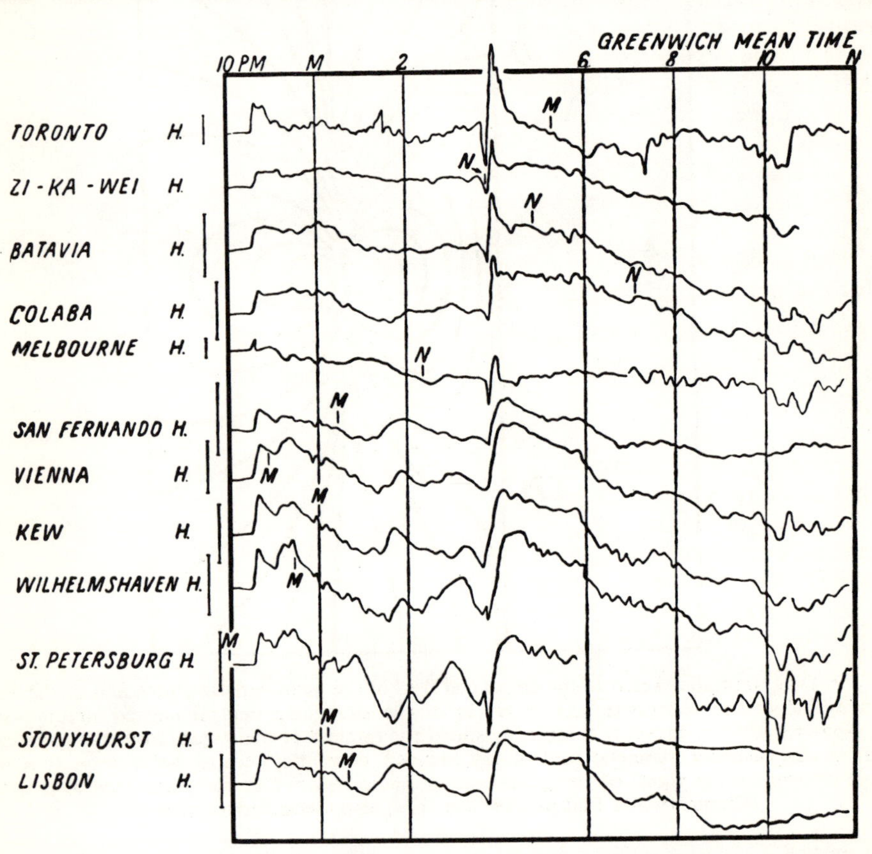

Fig. 3.3. Records of the variation of the horizontal magnetic force during the magnetic storm of 1885 June 24, 25. The scale of force varies from one station to another; it is indicated by the small vertical arrows, which in each case represent 100 γ or 0.001 gauss. The time interval represented by the distance between adjacent vertical lines is 2 hours. The letters *N* and *M* indicate the epochs of local noon and midnight.

might thus reach the earth's atmosphere, in two regions in high latitudes, one northern, the other southern, on the sunlit hemisphere. Chapman and Ferraro could not indicate the positions of these regions, because they could not determine the shape of the hollow and the form of the lines of force within it. Progress towards this has been made recently (see §§ 11–14). These horns have been discussed by various investigators (e.g., Mayaud 1955: Lebeau and Schlich 1962), in connection with the distribution of magnetic activity over the earth. But the importance and significance of the horns are still not clear.

3.3. The cylindrical sheet plasma problem

In 1940 Chapman and Ferraro gave the solution of a plasma problem that was intended to throw light on the nature of the motion of the particles of the solar gas when it approaches the geomagnetic field. The actual problem is very difficult; it is three-dimensional, and involves curved lines of magnetic force, and a compressible fluid. The illustrative plasma problem is fairly simply soluble mathematically, because in it the lines of magnetic force are straight, the particle motions are plane, there is symmetry about an axis, and the plasma is a surface not a volume distribution. Naturally the solution of this much simplified problem cannot illustrate all the features of the actual problem.

The plasma consists of positive ions and electrons, uniformly distributed over a cylindrical surface infinitely long in both directions, and of initial radius R_0. The total electric charge (e.s.u.) per unit length of the cylinder is Q_i for the ions, Q_e for the electrons; the whole is neutral, that is,

$$Q_i + Q_e = 0, \qquad Q_i = - Q_e = Q. \tag{3.2}$$

If N_i, N_e denote the number of ions and electrons respectively per unit length of the cylinder, and e_i, $-e$ their charges,

$$Q_i = N_i e_i, \qquad Q_e = - N_e e, \tag{3.3}$$

so that if the ions are singly charged $N_i = N_e$.

The axis of the cylinder is taken as the z axis of a right-handed system of cylindrical coordinates R, θ, z; the Oz direction will be called northward. Looking toward Oz with its north end upward, the westward direction round it is to the left, and the eastward direction is to the right, on the side of Oz nearer to the observer; θ is positive towards the east, and R increases outwards. Vector components along the R, θ, z directions may be indicated by corresponding subscripts to the symbols for the vectors.

A permanent magnetic field is present. Its lines of force are straight and upward, parallel to Oz. Its strength H_p at distance R from Oz is given by

$$H_p = M/R^n, \text{ where } n > 2. \tag{3.4}$$

In illustrative calculations we take $n = 3$; in this case the distribution of H_p in any plane z = constant is the same as for a dipole of moment M in its equatorial plane. This model field could be produced either by a suitable distribution of electric current, or by a suitable distribution of permanent magnetization. The cause, whatever it may be, is supposed unaffected by the plasma motion, and not to affect this motion except by its magnetic field H_p.

Chapman and Ferraro (1940) postulated that at time $t = 0$ all the particles are projected radially inward toward the axis Oz with the same speed $- U_0$. Chapman and Kendall (1961) gave the solution generalized for oblique symmetrical inward projection, in planes normal to Oz. In either case, each particle has no motion parallel to Oz, at any time.

The radial components of the forces on the particles will in general tend to separate the two sets of charges. When the charges all lie on the same sheet, the separative force may be denoted by eQ''/R. If $|Q''| < Q$, the two sheets will remain together, but as soon as $|Q''|$ exceeds Q, they will separate. The case resembles that of a force tending to raise a body from a table. So long as the force is less than the weight of the body, the body continues to rest on the table (though the interaction between it and the table is lessened); only when the force exceeds the weight does the body leave the table.

The R, θ components of the velocity $\mathbf{v}$ of a particle are denoted by U, V, with subscript i or e. By reason of the axial symmetry of the field, of the plasma, and of its initial projection, the charges of either kind always lie on a cylinder, of radius denoted by R_i or R_e; the radial and transverse velocity components U, V (with subscript i or e) are functions of t. The transverse motions constitute solenoidal currents of strength J_i, J_e (e.m.u.) per unit length of the cylinder, and

$$J_i = Q_i V_i/2\pi R_i c, \qquad J_e = Q_e V_e/2\pi R_e c, \qquad J = J_i + J_e. \tag{3.5}$$

Each solenoid produces a uniform field (H_i, H_e) within its sheet e.g.:

$$H_i = 4\pi J_i = 2Q_i V_i/R_i c, \qquad (R < R_i). \tag{3.6}$$

Outside the sheet ($R > R_i$), $H_i = 0$; likewise for H_e.

The total and partial fields $\mathbf{H}$, $\mathbf{H}_p$, $\mathbf{H}_i$, $\mathbf{H}_e$ are associated with corresponding vector potentials, which have no R and z components; they are related to the magnetic fields as in

$$\mathbf{H} = \text{curl}\, \mathbf{A}, \qquad H = \partial A/\partial R + A/R. \tag{3.7}$$

Thus

$$H = H_p + H_i + H_e, \qquad A = A_p + A_i + A_e, \qquad A_p = -M/(n-2)R^{n-1}, \tag{3.8}$$

$$A_i = 2\pi R J_i = Q_i V_i R/R_i c, \qquad (R < R_i), \qquad A_i = Q_i V_i R_i/Rc (R > R_i). \tag{3.9}$$

The radial gradient of A_i is discontinuous at $R = R_i$, but A_i itself is continuous there. Similar equations and remarks apply to A_e.

Once the particles have been set in motion, an electric field $\mathbf{E}$ arises, associated with $\mathbf{A}$ and a scalar potential φ as follows:

$$\mathbf{E} = -\partial\mathbf{A}/c\partial t - \text{grad}\, \varphi, \qquad \text{div}\, \mathbf{A} + \partial\varphi/c\partial t = 0. \tag{3.10}$$

As $\mathbf{A}$ has only a θ component, which is a function of R and t only, div $\mathbf{A} = 0$, so that $\partial\varphi/\partial t = 0$. Hence φ is a function of R only, independent of t, so long as no charged sheet sweeps over the point considered, causing a discontinuous change of φ. Thus

$$E_R = -\partial\varphi/\partial R, \qquad E_\theta = -\partial A/c\partial t. \tag{3.11}$$

At all times $E_z = 0$.

When $R_e \neq R_i$, $E_R = 0$ except between the two sheets ($R_i \gtrless R \gtrless R_e$); there $E_R = 2Q'/R$, where $Q' = Q_e$ or $Q' = Q_i$, according as the inner sheet is electronic or ionic. The discontinuity of E_R on crossing a charged sheet is due to the field of the local charges, which is of opposite sign on the two sides. The radial electric force on a charge in a surface distribution is due to the field of the non-local charges; this is the mean of the fields of the two sides. The radial force on the ions is eQ'/R_i; on the electrons it is $-eQ'/R_e$. Similarly the electromagnetic force on an ion is $e_i\mathbf{v}_i \times \mathbf{H}_i'/c$, where $\mathbf{H}_i'$ is the mean of $\mathbf{H}$ on the two sides of the sheet For ions and electrons respectively,

$$H' = H_p + H'_i + H_e, \qquad H' = H_p + H_i + H'_e, \qquad H'_i = \tfrac{1}{2}H_i, \qquad H'_e = \tfrac{1}{2}H_e. \tag{3.12}$$

When the two sheets are unseparated,

$$H' = H_p + H'_i + H'_e = H_p + \tfrac{1}{2}(H_i + H_e). \tag{3.13}$$

The equation of motion of an ion is

$$m_i\dot{\mathbf{v}}_i = e(\mathbf{E}'_i + \mathbf{v}_i \times \mathbf{H}'/c) \tag{3.14}$$

This has no z component. Its R and θ components are:

$$m_i(\dot{U}_i - V_i^2/R_i) = eQ'/R_i + eV_i(H_p + H'_i + H_e)_i/c \tag{3.15}$$

$$\frac{m_i}{R_i}\frac{d(R_iV_i)}{dt} = -\frac{e}{c}\frac{\partial}{\partial t}(A'_i + A_e)_i - \frac{eU_i}{c}(H_p + H'_i + H_e)_i. \tag{3.16}$$

Similar equations apply to the electrons, with the sign of e changed, and the accent applied to H_e instead of to H_i. The subscript to the bracket indicates the radius R_i or R_e at which the value of the bracket is to be taken. When the sheets are unseparated, the accent is to be given both to the ionic and electronic A and H.

Let D_i/Dt denote the "mobile operator" $\partial/\partial t + \mathbf{v}_i \cdot \text{grad}$, "following the motion of the (ionic) particle". Because A is a function of R and t only,

$$\begin{aligned}(D_i/Dt)(A_p + A'_i + A_e) &= (\partial/\partial t)(A'_i + A_e) + U_i(\partial/\partial R)(A_p + A'_i + A_e) = \\ &(\partial/\partial t)(A'_i + A_e) + U_i(H_p + H'_i + H_e)_i - (U_i/R_i)(A_p + A'_i + A_e)_i\end{aligned} \tag{3.17}$$

and

$$\begin{aligned}(D_i/Dt)\{R_i(A_p + A'_i + A_e)\} &= U_i(A_p + A'_i + A_e) + R_i(D_i/Dt)(A_p + A'_i + A_e) \\ &= R_i(\partial/\partial t)(A'_i + A_e)_i + R_iU_i(H_p + H'_i + H_e)_i.\end{aligned} \tag{3.18}$$

Consequently the θ equation of ionic motion can be written in the form:

$$(D_i/Dt)\{m_iR_iV_i + eR_i(A_p + A'_i + A_e)_i/c\} = 0. \tag{3.19}$$

Integrated, this gives

$$m_iR_iV_i + (eR_i/c)(A_p + A'_i + A_e)_i = m_iR_0V_{i0} + (eR_0/c)(A_{p0} + A_{i0} + A_{e0}) \tag{3.20}$$

where V_{i0} is the θ-component speed given to the ions at time $t = 0$, $R_i = R = R_0$. The corresponding equation for the electrons is

$$m_eR_eV_e - (eR_e/c)(A_p + A_i + A'_e)_e = m_eR_0V_{e0} - (eR_0/c)(A_{p0} + A_{i0} + A_{e0}) \tag{3.21}$$

3.4. The case of unseparated sheets

When the sheets are unseparated, so that $R_i = R_e = R$, in the R-component equations Q'' must replace Q', and in the integrals of the θ component equations both A_i and A_e must have accents; but in these integrals the accents for the A's may be omitted, as no space differentiation is involved, and the A's are continuous across the sheets. So long as the sheets remain together, so that also $U_i = U_e$, and $J = Q(V_i - V_e)/2\pi Rc$, the R and θ component equations for the ions and electrons may be combined in the following convenient ways, using the notations:

$$m = m_i + m_e, \qquad m' = m_i m_e/m, \qquad m'' = m' + eQ/c^2, \qquad m_i V_i + m_e V_e = mV,$$

$$V_i - V_e = V', \qquad m_i U_i + m_e U_e = mU, \qquad U_i - U_e = U', \tag{3.22}$$

so that

$$V_i = V + m_e V'/m, \quad V_e = V - m_i V'/m, \quad m_i V_i^2 + m_e V_e^2 = mV^2 + m'V'^2; \tag{3.23}$$

similarly for the U's (note that $U' = 0$ initially, and until separation occurs).

By adding the two R-component equations, we obtain

$$m\dot{U} = (mV^2 + m'V'^2)/R + (eV'/c)(H_p + H'_i + H'_e) \tag{3.24}$$

or, what is equivalent, by (6) and the corresponding equation for H_e, and by (13),

$$m\dot{U} = (mV^2 + m''V'^2)/R + eV'H_p/c. \tag{3.25}$$

Multiplying the ionic R-component equation by m_e/m and subtracting from it the electronic R-component equation multiplied by m_i/m, we obtain the following equation for Q'':

$$Q'' = m'(V_e^2 - V_i^2)/e - (R/mc)(m_e V_i + m_i V_e)(H_p + H'_i + H'_e); \tag{3.26}$$

this is because $U' = 0$ as long as the sheets stay together. Initially

$$Q'' = Q''_0 = - R_0 V_0 H_{p0}/c$$

if $V_{i0} = V_{e0} = V_0$ or $V'_0 = 0$, so that also

$$J_0 = 0, \qquad A_{i0} + A_{e0} = 0, \qquad H'_{i0} + H'_{e0} = 0.$$

If this value of Q''_0 exceeds Q numerically, the sheets will separate immediately on projection; the motion of the ions relative to the electrons will be outward if V_0 is negative, and inward if it is positive.

The θ-component ionic and electronic equations can similarly be combined, to give

$$VR = V_0 R_0, \tag{3.27}$$

$$m'V'R + (eR/c)(A_p + A_i + A_e) = m'V'_0 R_0 + (eR_0/c)(A_{p0} + A_{i0} + A_{e0}). \tag{3.28}$$

If initially $V' = 0$, so that also A_i and A_e are zero, the right-hand side in (28) reduces to $eR_0 A_{p0}/c$, and (28) may also be written:

$$m''V'R = (e/c)(R_0 A_{p0} - RA_p). \tag{3.29}$$

These equations give V, V', hence also V_i, V_e, as functions of R. Thus Q'' and $\dot{U}$ are obtained as functions of R; also, because $\dot{U} = \frac{1}{2}\, dU^2/dR$, we can obtain U as a function of R by integration, and by a further integration, R as a function of t. This completely solves the problem.

3.5. Radial projection: sheets unseparated

In their numerical discussion of the case of radial projection, Chapman and Ferraro took numerical values illustrative of the actual problem of solar streams projected towards the earth; thus R_0 was taken to be 1.5×10^{13} cm, and U_0 to be of order 10^8 cm sec. Values of Q were found that gave a field $H_i + H_e$ of order 10γ to 50γ, corresponding to the external part of the initial increase of the horizontal magnetic field of the earth in a magnetic storm. It was found that such values of Q were enormously greater than suffice to hold the ionic and electronic sheets together. Hence here, in describing the case of oblique projection, we deal only with the case when the sheets remain together. Also, following Chapman and Ferraro, the value of R_0 will here be taken to be infinity, because this much simplifies the mathematics, without disadvantage to the geophysical applications. Hence for the constant value of VR we write BU_0; B is the distance from Oz of the initial asymptote to the path of each particle. The angular momentum $NmRV$ of the sheet per unit length, and the distance B, are taken to be positive, if the angular momentum is eastward. The case considered by Chapman and Ferraro is $B = 0$. If the obliquity of projection is not the same for the two kinds of particles, $V_iR_i = B_iU_0$, $V_eR_e = B_eU_0$.

Because R_0A_{p0} is zero for projection from infinity (R_0 infinite),

$$V' = eM/(n-2)m''cR^{n-1}. \qquad (3.30)$$

For radial projection ($B = 0$, $V = 0$) this leads to the following expression for Q'':

$$Q'' = \frac{m_i - m_e}{m}\frac{(n-1)eM^2}{(n-2)^2m''c^2R^{2n-2}}. \qquad (3.31)$$

Clearly $Q'' = 0$ if $m_i = m_e$, so that in this case there is no tendency for the sheets of ions and electrons to separate from each other.

The "magnetic disturbance" field h, equal to $H_i + H_e$, is given by

$$h = 2QV'/Rc = 2eQM/(n-2)m''c^2R^n = 2eQH_p/(n-2)m''c^2, \qquad (3.32)$$

where H_p here denotes the value of H_p *at the position of the sheet at the time considered.* When Q is large, this approximates closely to $h = 2H_p$. (3.32a)

3.5a Equal oblique projection: sheets unseparated

The equation of radial motion, taking $U = U_0$ at $R = \infty$, can now be integrated and written thus:

$$(U/U_0)^2 = 1 - (B/R)^2 - (R_1/R)^{2n-2}, \quad R_i^{2n-2} = \{eM/(n-2)cU_0\}^2/mm''. \qquad (3.33)$$

Thus R_1 denotes the radius to which the sheet penetrates when $B = 0$, that is, when the projection is radial: because then $U = 0$ at $R = R_1$.

It is convenient at this point to make our equations non-dimensional by taking R_1 as the unit of length, and U_0 as the unit of speed. Thus our unit of time becomes R_1/U_0. Let

$$r = R/R_1, \qquad b = B/R_1, \qquad u = U/U_0, \qquad v' = V'/U_0. \qquad (3.34)$$

Our equations for V' and U can now be given the form:

$$v' = (m/m'')^{1/2}/r^{n-1}, \qquad u^2 = 1 - (b/r)^2 - 1/r^{2n-2}. \tag{3.35}$$

In considering the orbits of individual particles, in unseparated sheets ($R_i = R_e$), it is convenient to write

$$V_i = R_{ii}\,\dot\theta = R\dot\theta_i, \qquad V_e = R\dot\theta_e, \qquad V = R\dot\theta,\ V' = R\dot\theta', \tag{3.36}$$

so that

$$\dot\theta_i = \dot\theta + m_e\dot\theta'/m, \qquad \dot\theta_e = \dot\theta - m_i\dot\theta'/m. \tag{3.36a}$$

It is also convenient to reckon the coordinates θ from the radius to the position of the particle at the point of closest approach to the axis. The equations for V and V' in the new units can be thus expressed:

$$d\theta/dr = b/r^2u, \qquad d\theta'/dr = (m/m'')^{1/2}/r^n u. \tag{3.37}$$

3.6. The case n = 3

To make further progress in the solution it is now convenient to take $n = 3$. Then the equation for u can be solved in the form:

$$u^2 = (1 - 1/r^2 \tan\alpha)(1 + 1/r^2 \cot\alpha), \tag{3.38}$$

where

$$\alpha = \tfrac{1}{2}\arctan(2/b^2) \text{ or } b^2 = 2\cot 2\alpha. \tag{3.39}$$

The real root of the equation is

$$r^2 = r_2^2 = \cot\alpha = (1 + \tfrac{1}{4}b^4)^{1/2} + \tfrac{1}{2}b^2, \tag{3.40}$$

corresponding to the cessation of inward motion of the particles at the radius

$$R_2 = r_2R_1. \tag{3.41}$$

When B is large, (40) approximates to $r_2 = b$, indicating that the particle moves almost along its initial line of motion.

To determine θ as a function of r it is convenient to express r as follows, in terms of a new variable ϕ; the use of this symbol here need cause no confusion with its former meaning, of the scalar potential, in equations (7) to (11).

$$r = r_2\sec\phi = (\cot\alpha)^{1/2}\sec\phi. \tag{3.42}$$

Initially $\phi = -\tfrac{1}{2}\pi$, and finally $\phi = \tfrac{1}{2}\pi$. In terms of ϕ,

$$r^2u = \Delta\,\mathrm{cosec}\,\alpha\tan\phi\sec\phi, \qquad dr = (\cot\alpha)^{1/2}\sec\phi\tan\phi\,d\phi, \tag{3.43}$$

where

$$\Delta^2 = 1 - \sin^2\alpha\sin^2\phi = 1 - w, \text{ where } w = \sin^2\alpha\sin^2\phi. \tag{3.44}$$

Thus the equation for $d\theta/dr$ is equivalent to

$$d\theta/d\phi = (\cos 2\alpha)^{1/2}/\Delta. \tag{3.45}$$

Hence

$$\theta = (\cos 2\alpha)^{1/2} F(\alpha, \phi), \tag{3.46}$$

in the notation of Jahnke and Emde, so that F denotes the incomplete elliptic integral of the first kind. The initial and final values of θ are $-\theta_2$ and θ_2, where

$$\theta_2 = (\cos 2\alpha)^{1/2} K(\sin \alpha). \tag{3.47}$$

The signs of B, b, α, K, θ and θ_2 are all the same. When $B = 0$, $\theta_2 = 0$.

In the present notation, the equation for $d\theta'/dr$ may be written

$$d\theta'/d\phi = (m/m'')^{1/2}(\sin \alpha \cos \phi)/\Delta \tag{3.48}$$

On integration, this gives

$$\theta' = (m/m'')^{1/2} \arcsin (\sin \alpha \sin \phi). \tag{3.49}$$

Initially

$$\phi = -\tfrac{1}{2}\pi \text{ and } \theta' = -\theta'_2, \tag{3.50}$$

where

$$\theta'_2 = (m/m'')^{1/2}|\alpha|, \tag{3.51}$$

and $|\alpha|$ denotes the positive numerical value of α. When the initial projection is radial, so that $B = 0$ and $\alpha = \frac{1}{4}\pi$, θ' has the value θ_1' given by

$$\theta'_1 = \tfrac{1}{4}(m/m'')^{1/2}\pi. \tag{3.52}$$

Hence for any other value of B or b,

$$\theta'_2 = (4|\alpha|/\pi)\theta'_1. \tag{3.53}$$

The larger the value of B (the more oblique the projection), and consequently the smaller the value of α, the smaller is the difference between the azimuthal changes of θ_i and θ_e. When $b = 0$, $\alpha = \frac{1}{4}\pi$; when b is small, α is approximately $\frac{1}{4}(\pi - b^2)$, and when b is large, α is approximately $1/b^2$. We may note that

$$\sin \alpha = 2^{-1/2}[1 - 1/(1 + 4b^4)^{1/2}]^{1/2}, \qquad \sec 2\alpha = (1 + 4b^4)^{1/2}. \tag{3.54}$$

The total deflections of the velocity vector of an ion or electron will be denoted by χ_{i2} and χ_{e2}. They are given by

$$\chi_{i2} = \pm (\pi - 2\theta_2) - 2(m_e/m)\theta'_2, \qquad \chi_{e2} = \pm (\pi - 2\theta_2) + 2(m_i/m)\theta'_2. \tag{3.55}$$

The positive sign is here to be taken when B (and hence also θ_2) is positive, and the negative sign when B is negative, *except* when B is so small that θ'_2 (which is always positive) numerically exceeds $m\theta_2/m_e$ (in this case the positive sign is to be taken in the expression for χ_{i2}), or when θ'_2 numerically exceeds $m\theta_2/m_i$ (in this case the negative sign is to be taken in the expression for χ_{e2}). In these special cases the orbits of the particles are looped.

For radial projection, $\theta = 0$, and

$$\theta' = \pm \tfrac{1}{2}(m/m'')^{1/2}\{\tfrac{1}{2}\pi - \arcsin (1/r^2)\}. \tag{3.56}$$

At the minimum distance, $r = 1$, $\theta' = 0$; the whole variation of θ', from infinity to the minimum distance and back to infinity, is $\frac{1}{2}\pi(m/m'')^{1/2}$; for the ions the radius from the axis sweeps through an angle less than this in the ratio m_e/m, and for the electrons, in the ratio m_i/m; the sweep is eastwards (positive) for the ions, westwards (negative) for the electrons.

For radial projection, the greatest value (Q''_1) of Q'' occurs at the minimum distance R_1, where its value is given (for any value of n) by

$$Q''_1 = (n - 1)(m_i - m_e) U_0^2/e, \tag{3.57}$$

or $Q''_1 = 2(m_i - m_e)U_0^2/e$ for $n = 3$. Thus no separation will occur (in the case of radial projection, $n = 3$) if $Q > 2(m_i - m_e)U_0^2/e$. For $U_0 = 10^8$ cm/sec., this requires that $Q > 67$ esu if the ions are protons.

The values of Q needed to produce appreciable disturbing fields, when the numerical magnitudes inserted in the formulae are analogous to those involved in the production of geomagnetic storms, are immensely greater than this, being of order 10^{10}. Hence in m'' the term m' is quite negligible in comparison with eQ/c^2; $c^2m'/e = 1700$, so that if Q is large compared with 1700 esu we may substitute eQ/c^2 for m'' in our formulae. To produce a maximum disturbing field h_1 (at $R = R_1$) with radial projection, we require

$$Q = (e/m U_0^2)(\tfrac{1}{4}Mh_1^2)^{2/3}. \tag{3.58}$$

Taking $U_0 = 10^8$ cm/sec., as before, and $M = 8.06 \times 10^{25}$ cm³ gauss, for $h_1 = 10^{-4}$ and $h_1 = 10^{-3}$ (that is, h_1 is 10γ or 100γ), Q has the respective values 0.99×10^{10} and 2.12×10^{11}.

In terms of the same "observables" U_0 and h_1, with the same neglect of m' in comparison with eQ/c^2, we have, for radial projection,

$$R_1 = (2M/h_1)^{1/3}, \qquad R_1/a = (2H_0/h_1)^{1/3}, \tag{3.59}$$

where a denotes the earth's radius, and H_0 the dipole intensity at the earth's surface at the equator.

For radial projection the change in θ_e, from infinity to infinity, can be expressed as follows, neglecting m' in comparison with eQ/c^2:

$$-\pi mc U_0/(2Mh_1^2)^{1/3}e \text{ radians or } 3.45 \times 10^{-11} U_0/h_1^{2/3} \text{ degrees}; \tag{3.60}$$

the minus sign indicates that the change of azimuth is westward.

The following Table gives some numerical particulars for the case of radial projection, using the above values of M and U_0, for two values of h_1.

TABLE 3.1

Radial projection, $M = 8.06 \times 10^{25}$, $U_0 = 10^8$, $a = 6.37 \times 10^8$

	$h_1 = 20\gamma$	$h_1 = 200\gamma$
(1) R_1/a (distance of closest approach; a = earth radius)	14.6	6.8
(2) Q (e.s.u.; charge per cm length of cylinder)	2.5×10^{10}	5.3×10^{11}
(3) N (number of ions or electrons per cm length)	5.2×10^{19}	1.1×10^{21}
(4) ν_1 (number/cm² at minimum distance)	8.9×10^8	4.1×10^{10}
(5) Total change of electron azimuth, $2\theta_{e\,\infty}$	$-1°.01$	$-0°.22$
(6) Total change of ion azimuth, $2\theta_{i\,\infty}$	5.5×10^{-3}	1.2×10^{-3}
(7) J_1, the maximum current (amps) per cm length	1.6×10^{-4}	1.6×10^{-3}
(8) $2J_1a$, maximum current (amps) for length = earth diameter	2×10^5	2×10^6
(9) Time unit R_1/U_0 in seconds	93	43

The aspects of magnetic storm theory specially illuminated by the solution of this plasma problem relate only to the initial phase of the storm; they are:

(i) the order of magnitude of the distance at which solar gas is likely first to be appreciably affected by the geomagnetic field (this aspect had already been illustrated by the plane rigid sheet problem); see row 1 of the above Table;

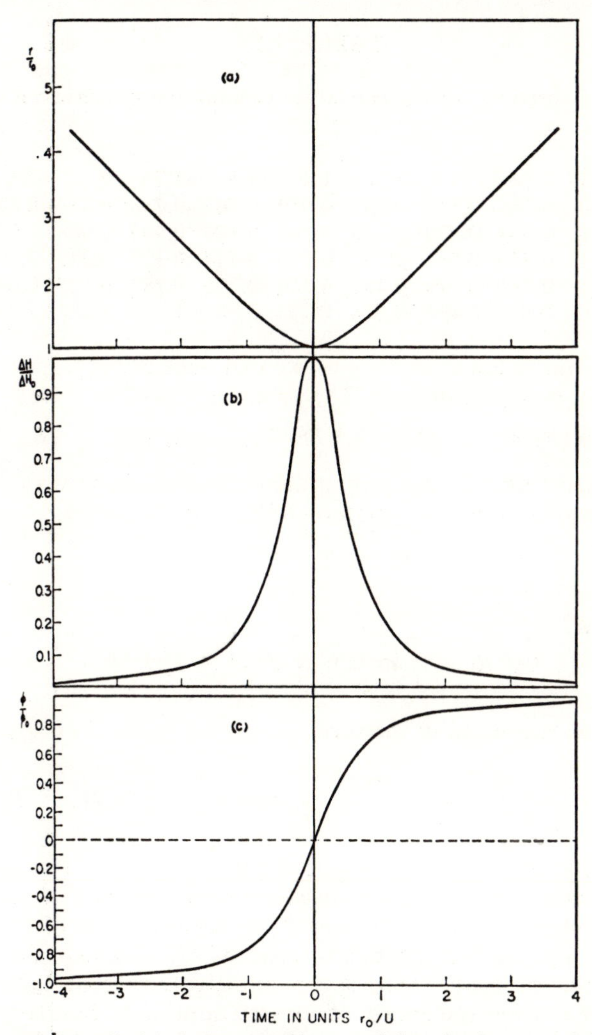

Fig. 3.4. To illustrate the cylindrical sheet problem, radial projection, $n = 3$ (§3.6 and Table 3.2): (a) R/R_1 as a function of time, measured in the unit R_1/U_0; (b) h/h_1; (c) θ'/θ'_0. (Chapman, 1960, p. 927.)

(ii) how the electrostatic force between the opposite charges in the gas keeps the ions and electrons travelling together, enabling the electrons to penetrate the field far more deeply than they could if travelling alone;

(iii) the nearly complete turning back of the particles that most directly approach the earth; see rows 5, 6;

(iv) the order of magnitude of the electric currents involved; see rows 7, 8;

(v) the time of growth of the field h to its maximum value.

This last aspect and others are illustrated by the following Table and by Fig. 4.

TABLE 3.2

Radial projection, for $eQ/m'c$ large; general non-dimensional solution

R/R_1	1	1.1	1.2	1.3	1.4	1.5	1.75	2	2.5	5.0
θ'/θ'_∞	0	0.381	0.511	0.597	0.659	0.707	0.788	0.839	0.898	0.975
U/U_0	0	0.563	0.720	0.806	0.860	0.896	0.945	0.968	0.987	0.999
V'/V'_1	1	0.826	0.694	0.592	0.510	0.444	0.327	0.250	0.160	0.040
h/h_1	1	0.751	0.579	0.455	0.364	0.296	0.187	0.125	0.064	0.008
U_0t/R_1	0	0.329	0.484	0.614	0.733	0.847	1.118	1.380	1.890	4.400

This Table is non-dimensional so long as we can neglect the term $m'c^2/eQ$ in m'', as in all cases of geomagnetic interest. The values in rows 2 to 5 are:

$$1 - (2/\pi)\arc\sin(R_1/R), \qquad (1 - R_1^4/R^4)^{1/2}, \qquad (R_1/R)^2, \qquad (R_1/R)^3. \tag{3.61}$$

The last column gives time from the minimum distance, in terms of the time unit R_1/U_0. It is obtained thus as a function of R/R_1 or r instead of

$$dt = dR/U = (R_1/U_0)\,dr/u, \tag{3.62}$$

so that

$$U_0t/R_1 = \int dr/u = \int dr/(1 - 1/r^4)^{1/2}. \tag{3.63}$$

This integral can be expressed in the form

$$1 - R_1/R + \frac{1}{2^{3/2}}\int_x^{1/2} \frac{dx}{x^{1/4}(1-x)^{5/4}}, \text{ where } x = \tfrac{1}{2} - \tfrac{1}{2}(1 - R_1^4/R^4)^{1/2} \tag{3.64}$$

Thus when $R = R_1$, $x = \frac{1}{4}$, and when R is infinite, $x = 0$. The integral can be calculated by expanding the integrand in a series of powers of x, and integrating term by term. Fig. 4 plots R/R_1, h/h_1, and θ'/θ'_∞ as a function of time, in terms of the time unit R_1/U_0 given, for particular values of M, U_0 and h_1, in the last row of Table 1. Most of the change of radial speed, namely 97 percent, takes place in the 2 time units before or after the epoch of closest approach. Similarly most of the change in azimuth of the radius vector to any particle, namely 84 percent, takes place in the same interval of time; and in this interval the lateral speeds, which give the electric current, complete the last 75 percent of their rise to their maximum value, at minimum distance. The time unit, in both the numerical cases considered, is so small that these results show how rapid is the development of the important changes in the system, when (but only when) the particles are near their points of closest approach. It is of interest to compare the distance R_1 of closest approach of the particles to the field axis with the distances of closest approach of a single proton or electron similarly projected radially towards the axis, in the absence of any companion particles. These distances are given by the formulae developed by Störmer (1955) for the motion of a single charged particle in a magnetic dipole field. In the equatorial plane of the dipole the field distribution of magnetic intensity is the same as for the field H_p above considered, if $n = 3$. Störmer's formulae show that a

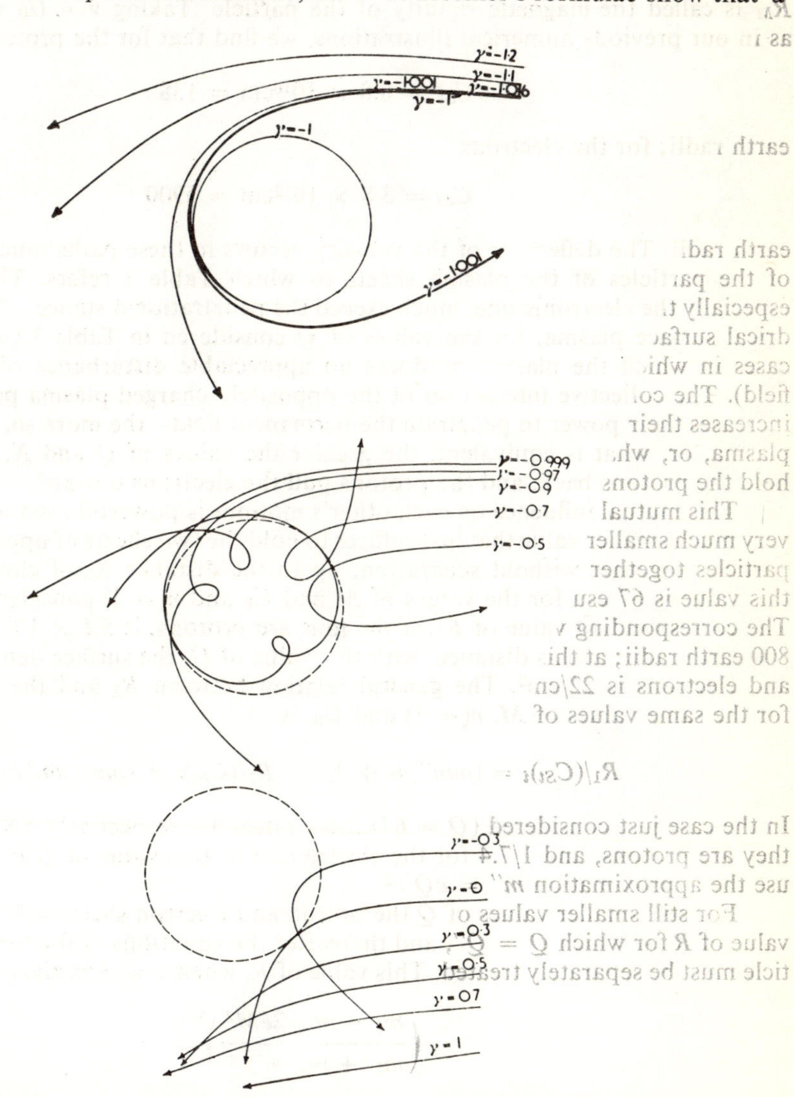

Fig. 3.5. Types of orbit of particles in the equatorial plane of a dipole field; the circle is the orbit with radius unity, in terms of the Störmer length unit. (Störmer, 1955, p. 221.)

particle projected with speed v in this plane, radially towards the axis, follows a simple path (Fig. 5), and attains the distance C_{St} from the axis (this case corresponds to zero angular momentum of the particle about the axis, so that Störmer's integration constant γ is zero). Here

$$C_{St} = (Me/mvc)^{1/2} = (H_0a^3/R_M)^{1/2}, \qquad R_M = mvc/e; \tag{3.65}$$

R_M is called the magnetic rigidity of the particle. Taking $v = U_0 = 10^8$ cm/sec., as in our previous numerical illustrations, we find that for the protons

$$C_{St} = 8.8 \times 10^{10}\text{cm} = 138$$

earth radii; for the electrons

$$C_{St} = 3.8 \times 10^{12}\text{cm} = 5900$$

earth radii. The deflection of the velocity vectors in these paths much exceeds that of the particles of the plasma sheets to which Table 1 refers. These distances, especially the electronic one, much exceed the penetration distances R_1 for the cylindrical surface plasma, for the values of Q considered in Table 1 (which relate to cases in which the plasma produces an appreciable disturbance of the magnetic field). The collective interaction of the oppositely charged plasma particles greatly increases their power to penetrate the permanent field—the more so, the denser the plasma, or, what is equivalent, the greater the values of Q and N. The electrons hold the protons back, and the protons pull the electrons onward.

This mutual influence on each other's motions is powerful even when Q has the very much smaller value that just suffices to hold the two sheets of oppositely charged particles together without separation, up to the distance R_1 of closest approach; this value is 67 esu for the values of M and U_0 and $n(= 3)$ considered in Table 1. The corresponding value of R_1, if the ions are protons, is 5.1×10^{11} cm, or about 800 earth radii; at this distance, with this value of Q, the surface density of protons and electrons is $22/\text{cm}^2$. The general relation between R_1 and the values of C_{St}, for the same values of M, $n(= 3)$ and U_0, is

$$R_1/(C_{St})_i = (mm''/m_i^2)^{1/4}, \qquad R_1/(C_{St})_e = (mm''/m_e^2)^{1/4}. \tag{3.66}$$

In the case just considered ($Q = 67$) these ratios are respectively 5.8 for the ions if they are protons, and 1/7.4 for the electrons. For this value of Q it is not valid to use the approximation $m'' = eQ/c^2$.

For still smaller values of Q the proton and electron sheets will separate at the value of R for which $Q = Q''$, and thereafter the equations of the two kinds of particle must be separately treated. This value of R, when $n = 3$, is given thus by (31):

$$\left(\frac{m_i - m_e}{m_1 + m_e}\frac{2eM^2}{m''c^2}\right)^{1/4}. \tag{3.67}$$

Using the notation

$$m'_i = m_i + eQ/c^2, \qquad m'_e = m_e + eQ/c^2, \tag{3.68}$$

the equations of lateral motion for radial projection from infinity then take the form:

$$m'_i V_i - (eQ/c^2)(R_i/R_e) V_e = eM/(n-2)cR_i^{n-1}, \tag{3.69}$$

$$m'_e V_e - (eQ/c^2)(R_e/R_i) V_i = -eM/(n-2)cR_e^{n-1}. \tag{3.70}$$

In these equations the addition of eQ/c^2 to m_i and m_e on the left comes from the terms that involve A'_1 and A'_e; they represent an increase of virtual azimuthal inertia of the particles owing to self-induction; the second term on the left of these equations represents the effect of mutual induction between the two sheets. The same influence of self-induction increasing the resistance to change of relative azimuthal velocity before the sheets separate is represented in the equation (29) of combined angular momentum. For the above case of "just-no-separation" ($Q = 67$) the fractional change in the mass-coefficient m'' on the left, which for $Q = 0$ is

$$m_i m_e/(m_i + m_e)$$

or approximately m_e, is 67/1700 or about 4 percent.

As Q tends to zero, the motions tend to the Störmerian forms, as the mutual influence of the particles declines.

The equations of radial motion after separation can easily be written down, and are given by Chapman and Ferraro (1940), but they need not be considered here. During the separated motion there are four unknowns to be determined, instead of (effectively) only two as in the combined motion. After separation (in the case of initially radial projection) the electrons lag behind the ions; a curious result mentioned by Chapman and Ferraro as a probability is that if the lag becomes sufficiently great, the lateral speed V_e of the electrons can be reversed, becoming eastward.

3.7. Energy transformation: radial projection from infinity without separation

The energy density of the uniform field h of magnetic disturbance produced at any time by the (combined) plasma sheets is $h^2/8\pi$. Its integrated amount E per unit length of the cylinder when the radius of the sheet is R is $h^2R^2/8$. Taking $n = 3$. this is given by

$$E = \tfrac{1}{2}(eQM/m''c^2R^2)^2. \tag{3.71}$$

The loss of kinetic energy of radial motion per unit length of the sheet is

$$\tfrac{1}{2}Nm(U_0^2 - U^2).$$

By (33) this equals

$$\tfrac{1}{2}NmU_0^2(R_1/R)^4$$

for radial projection ($B = 0$); and this is equal to

$$\tfrac{1}{2}\frac{Ne^2M^2}{m''c^2R^4}. \tag{3.72}$$

This loss exceeds the magnetic energy E; the difference is made up by the kinetic energy of the transverse motion, which is $\frac{1}{2}Nm'V'^2$ (by (23)), since $V = 0$ for radial projection); and by (30) this transverse energy E' is

$$\tfrac{1}{2}Nm'e^2M^2/m''^2c^2R^4. \tag{3.73}$$

The difference between the loss of radial kinetic energy and the gain of transverse energy is easily shown to equal E.

The ratio E'/E is given by

$$E'/E = m'c^2/eQ; \tag{3.74}$$

this is constant throughout the motion, and independent of the initial speed of radial projection. It is extremely small for sheets that are able to produce an appreciable magnetic disturbance; its value is $1700/Q$, and for the two cases illustrated in Table 4 this has the values 6.8×10^{-8} and 3.2×10^{-9}. Thus the conversion of radial kinetic energy into magnetic energy is remarkably efficient for sheets that can cause appreciable magnetic disturbance. When the radial motion ceases, at the radius R_1, the initial kinetic energy of the plasma sheet is almost entirely converted into magnetic energy, and the small remaining kinetic energy is almost all possessed by the electrons. That of the ions is smaller in the ratio m_e/m_i. Even for the electrons the energy is extremely small; for example, in the first case illustrated in Table 4, it is 6.3×10^{-33} erg per electron; for all the particles per unit length of the plasma sheet it is 3.3×10^{-3} erg.

However, for sheets for which Q is comparable with or less than 1700 esu, the interaction with the permanent field only slightly changes the total kinetic energy of radial and lateral motion. The proportion transformed into field energy (in all cases, whatever the value of Q) is

$$eQ/m''c^2 \text{ or } Q/1700 \text{ (esu)}.$$

Thus for $Q = 67$ esu, just sufficient, in the case above referred to, to prevent separation of the sheet into two, the change of energy from kinetic to magnetic is only 4 percent. As Q tends to zero, the trend is to the state of constant energy for each particle, as in the problem studied by Störmer.

For radial projection the lateral speed v' at minimum distance ($r = 1$) is $(m/m'')^{1/2}$, by (35). Hence V_e/U_0 at this distance, for such values of Q as are considered in Table 4 (namely values much exceeding 1700 esu), is very approximately $(1700m_i/m_eQ)^{1/2}$, or $1770/Q^{1/2}$. Consequently, for the cases considered in Table 1, the maximum lateral speed of the electrons is very small compared with their initial speed U_0.

The field of the plasma current is superposed on the permanent field H_p. As the two fields are in the same direction, the total field in the space within the sheet is $H + h$. The magnetic energy density there is $(H + h)^2/8\pi$; the change of energy density is $h^2/8\pi + Hh/4\pi$. The conservation of energy previously mentioned takes no account of the latter term. The case may be compared with the energy of the field of two currents J_1, J_2, whose self and mutual inductions are L_{11}, L_{12}, L_{22}; the field energy over the whole space is

$$\tfrac{1}{2}(L_{11}J_1^2 + 2L_{12}J_1J_2 + L_{22}J_2^2).$$

The initial postulate, in the problem here considered, that the system of currents (or permanent magnetism) that produces the permanent field is uninfluenced by the induced currents, implies that L_{12} must be reckoned zero in the energy calculation. In any case, the motion of the particles is influenced only by the field at the location of the sheet, and does not depend on the total field, and its energy, in the space within the sheet.

Though the analysis of this system and its changes throws light on the phenomena of impact of an ionized gas upon a magnetic field, it does not illustrate the shielding of the field in the gas behind the surface of the hollow carved by the geomagnetic field in a stream or shell of solar gas.

On account of the difference between the geometry of the cylindrical plasma sheet here considered, and that of the rigid perfectly conducting plane sheet previously considered, there is a difference between the distances of closest approach required to produce a given perturbation of the permanent field—namely 5.8 earth radii for $h = 20\gamma$ in the plane case, and 14.6 earth radii in the cylindrical plasma problem. Such a difference was to be expected; for a plane plasma sheet approaching the geomagnetic dipole field, the value of the distance of closest approach is likely to lie between the two values.

3.8. Oblique projection from infinity

In the case of radial projection of the particles, the ions and electrons are deflected to right and left respectively, but the angles of deflection are very small, in the cases of geomagnetic interest, such as Table 1 illustrates. The azimuthal angles of the particles of the two kinds are in this case $m_e\theta'/m$ for the ions, and $-m_i\theta'/m$ for the electrons; see (36a); for radial projection V and θ are always zero. When the projection is oblique (Q being supposed great enough to prevent separation), θ' is smaller than for radial projection, so that the difference between the electron and ion paths is even less than that shown in Table 1. But the deflection common to both particles, represented by the change in θ, which is no longer zero, is much greater than in radial projection, except when the projection is very nearly radial. Using the equations (52), (53), the values of θ'_2 and θ'_2/θ'_1 have been calculated for various values of b (Chapman and Kendall 1961), together with the corresponding values of α, r_2, and h_2/h_1. Fig. 6 illustrates the values graphically. The Figure also gives values of r_3, defined as the radius of a rigid elastic cylinder (without any field of force) that would deflect the particles from their original to their final asymptotic lines of motion, by elastic rebound. In this imagined change of motion the particles travel in straight lines until they collide with the cylinder; then they are suddenly reflected so as to move along the final asymptote of the (real) motion.

In calculating r_3, only the common part of the deflection is considered; then it is easy to see that $r_3 = b/\sin\theta_2$. As b tends to zero, r_3 tends to $2^{1/2}/K(\pi/4)$ or 0.765. As b tends to infinity, r_2/b and r_3/b both tend to unity.

Fig. 7 shows the form of the paths of the individual particles (ignoring θ') for different values of b. Each path corresponds to a different case of a plasma cylinder, all of whose particles perform the same form of path, with all orientations about the axis. These paths for oblique projection may be contrasted with those followed by *single* particles similarly projected in the same field. As calculated and drawn by Störmer (1955), these are shown in Fig. 5, for various values of the integration constant γ. The curves apply to negatively or positively charged particles, according as the earth-dipole equatorial plane of their motion is supposed to be viewed from its northern or southern side, respectively. Störmer's parameter γ is one-half the distance of the line of initial projection—from infinity—from the

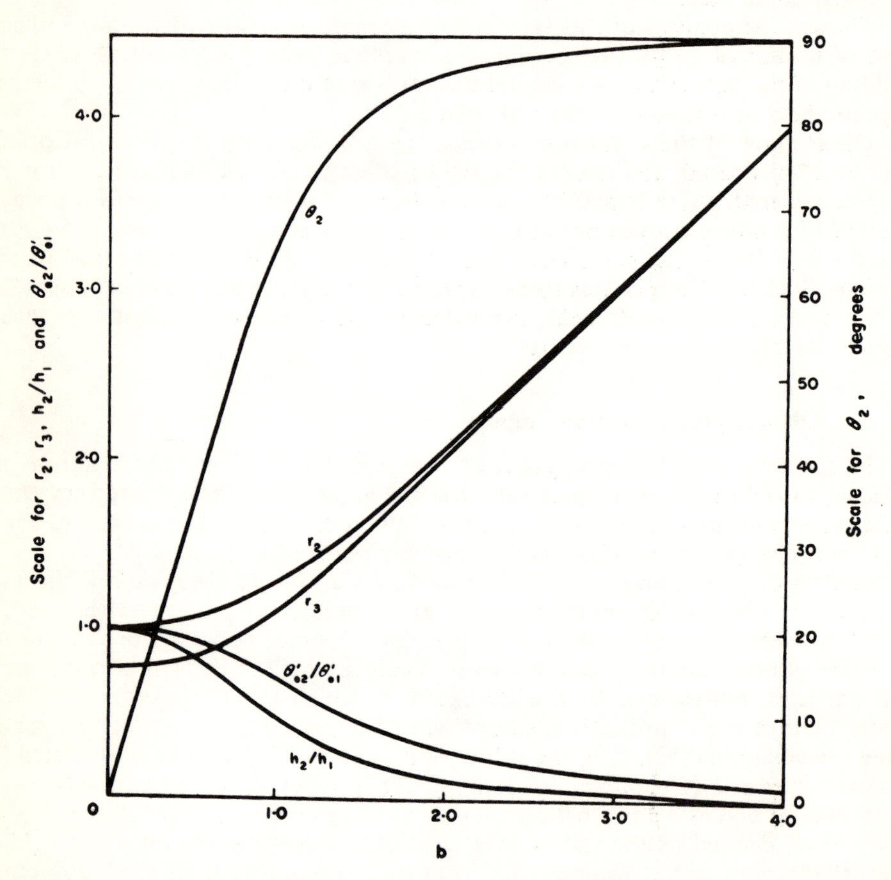

Fig. 3.6. To illustrate the cylindrical sheet problem: oblique projection, $n = 3$ (§3.8 and Table 3.6). r_2, r_3, $\theta'_{e2}/\theta'_{e1}$, h_2/h_1 and θ_2, as functions of b, the angular momentum, (Chapman and Kendall, 1961, p. 150.)

dipole axis, reckoned in Störmer length units C_{St}. For positive values of γ the Störmerian paths are simple; these values correspond to projection from one side of the dipole plane of radial projection (the left side for electrons, the right side for protons); the range of azimuthal angle, and the distance of closest approach, steadily increase with γ, from $\gamma = 0$. For negative values of γ, numerically increasing, the change of azimuthal angle decreases, and becomes negative near $\gamma = -0.5$, when the paths become looped; at first the loop overlaps the circle of unit radius.

then it comes to lie wholly within this circle. From $\gamma = -1$ onwards the paths again become simple; the range of azimuthal angle decreases, and the distance of closest approach steadily increases, as $-\gamma$ tends to infinity. This behavior partly agrees with, and partly differs from, that of the particles of a plasma sheet, as illustrated in Fig. 7.

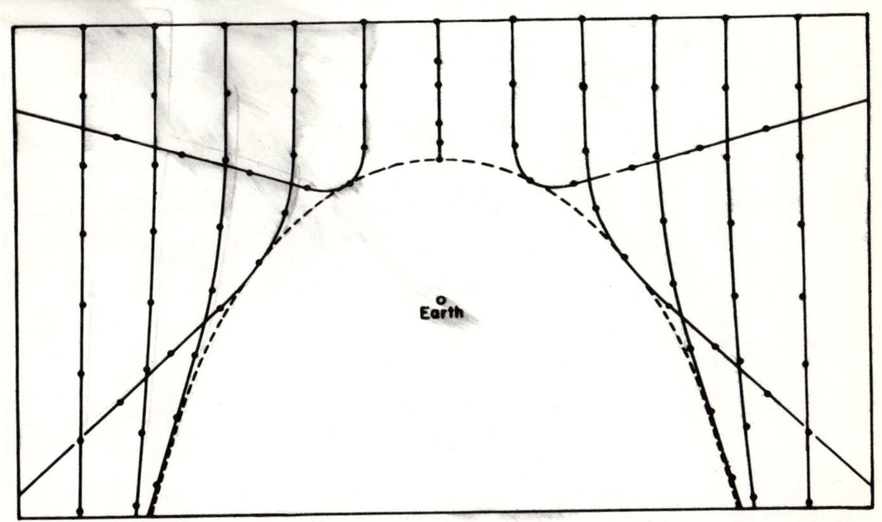

Fig. 3.7. The trajectories of individual particles of an initially plane ionized sheet projected towards a "cylindrical" magnetic field. The broken line represents the envelope of the paths. (Chapman and Kendall, 1961, newly drawn.)

It is tempting but unreliable to take the *envelope* of the paths in Fig. 7, with all their initial directions parallel, to be the form of the hollow that would be produced in a continuous *stream* of particles, projected from infinity towards the "cylindrical" magnetic field. The two physical situations are, however, too different for this to be valid.

Alternatively the paths shown in Fig. 7 may be used to construct the successive forms of a plane plasma sheet projected perpendicularly to itself towards the "cylindrical" field; each particle is taken to follow the same path as it would if it were a member of a cylindrical plasma sheet all of whose particles are projected with the same value of b. The result is shown in Fig. 8. The sides of the sheet proceed onward almost unaffected by the field. The central part is reflected away from the field after attaining a certain minimum distance, so that it expands indefinitely in a somewhat oval form. The picture does not accurately represent what would happen to such a sheet, for reasons discussed in detail by Chapman and Kendall (1961); the representation is likely to be qualitatively correct, but the currents flowing in the sheet when deformed have not been calculated, nor their magnetic field.

The oppositely charged particles of a neutral plasma *line*, initially straight and perpendicular to a dipole, projected towards the dipole perpendicularly to the line, will thereafter follow plane paths; those directed towards the dipole itself will be turned back with opposite deflections; those whose initial lines of motion cross the dipole axis on either side of the dipole will be curved somewhat as in Fig. 7. The addition to the virtual mass, owing to self and mutual induction, will be smaller in this case than in that of the cylindrical plasma sheet. The problem of their motion

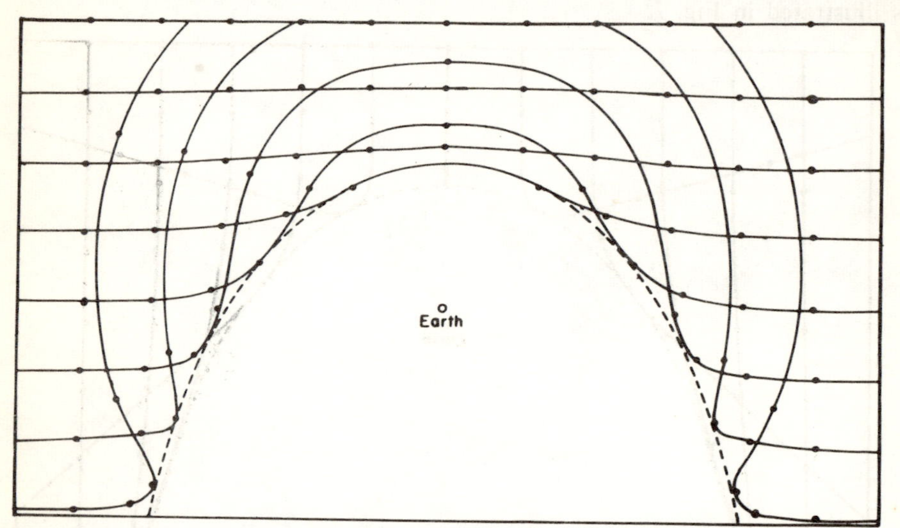

Fig. 3.8. The successive shapes of an initially plane ionized sheet projected towards a "cylindrical" magnetic field. The broken line represents the envelope of the paths of individual particles. (Chapman and Kendall, 1961, p. 154.)

has probably never been investigated; it is difficult, because the current along the line will vary, and electric charge and field will be set up. It must also be very difficult to determine the motion of a plane neutral plasma sheet projected towards the field of a dipole.

3.9. The impact of a plane-faced volume stream upon a plane-stratified field

The magnetic disturbance produced by any kind of plasma *sheet* impinging on the geomagnetic field will be brief. Very brief disturbances are often shown by the magnetic records. Some of them may well be the result of the impact of a relatively thin plasma sheet of some form, though of course not one that is merely a "surface" sheet. But the increase of the horizontal force at the earth's surface, which is a usual feature—the first phase—of magnetic storms, may persist for hours. The above study of the impact of a *sheet* upon a magnetic field suggests that a *volume* of plasma flowing to and around the earth is responsible for the first phase of a magnetic storm. This might be studied by an extension of the above analysis to the case of a succession of cylindrical plasma sheets projected directly or obliquely towards a cylindrical field. Such a succession of sheets could to some extent simulate a volume distribution of plasma. This problem is difficult, and has not been investigated in any detail, nor has the still more difficult problem of a thick cylindrical shell of plasma (both problems were mentioned in the 1940 paper summarized above. cf. p. 266).

Instead, another form of idealization of the actual solar plasma problem has been studied, in which the permanent magnetic field is undirectional (along Oy) and varies in intensity H_p only in one direction (z), as follows:

$$H_p = H_0(a/z)^n = M/z^n, \qquad n > 2, \qquad z > 0. \qquad (3.75)$$

The same remarks apply to this field as are made at the end of the paragraph containing equation (4).

The vector potential A_p is given by

$$A_p = - M/(n - 1)z^{n-1}. \qquad (3.76)$$

In the following discussion only the value $n = 3$ will be considered, though corresponding results are obtainable for other values of $n(> 2)$. Thus (75), (76) will be written

$$H_p = M/z^3, \qquad A_p = - M/2z^2, \qquad z > 0. \qquad (3.77)$$

Such a magnetic field, varying only in one direction instead of in two (as for the "cylindrical" field discussed above), is a step further away from the real geomagnetic field, which varies in all three (x, y, z) directions.

It would be simple, and of interest, to work out the problem of the projection, into this field, of a uniform neutral plasma plane *sheet*, whose z coordinate is a function of time. This would be analogous to the problem of the cylindrical plasma sheet already discussed.

Such a sheet, thus projected from $z = + \infty$ into the "plane" magnetic field (77), will develop electric current along the x direction. This will add a uniform field, also along Oz, to the permanent field, in the space in front of the sheet, and an equal and opposite field in the space already traversed by the sheet. In the latter region, as the "disturbance" field is uniform and the permanent field tends to zero, the latter is more than shielded—it is reversed beyond a certain value of z. This does not affect the solution for the plasma sheet, but such a feature adds unreality to the more general (and more difficult) problem of a *volume* flow of initially uniform neutral plasma, with a plane face ($z = z_0$, where z_0 is a function of t), into the permanent field.

The latter problem has been discussed by Ferraro (1952). It has not been possible to solve it with the simple rigor of the solution of the cylindrical sheet plasma problem. The volume flow problem was first discussed in a more general form (for impact upon the earth's dipole field) by Chapman and Ferraro 1931; 1932). Ferraro's 1952 discussion, like the 1940 cylindrical plasma sheet investigation, removed some misconceptions in the earlier work. In other respects his improved treatment largely confirmed some of the earlier findings, particularly as regards the time of development of the disturbing field, while the stream front is being halted by the permanent field. In part of his study he treated the volume stream as a succession of plane sheets. Those in front are the most retarded; they are overtaken by sheets to the rear, previously partly shielded from the permanent field by the currents induced in the gas further forward. The foremost sheets successively retreat into the stream and continue to move backward. Ferraro concluded (his p. 30) that the density at the front of the stream, at its deepest penetration into the permanent field, would not be largely increased, though he did not calculate its value (p. 47); this corrected the original Chapman-Ferraro expectation of much heaping up of gas, with greatly increased density, at the head of the stream. At that time the relief provided by the reversal of the forward motion was not understood.

The protons penetrate a little deeper into the permanent field than do the electrons. Thus the stream front is polarized normally to itself, and is the seat of an electric field in the z direction. This polarization and its electric field, like the induced electric currents, occupy only a thin layer near the stream front.

Ferraro's analysis is not reproduced here; only some of the main results and formulae (with page references to his paper) are quoted. His problem was time-dependent, because the advance of the stream into the field was considered. Perhaps a more rigorous solution could be obtained for a long-continuing stream, when a steady state of motion had been achieved. As in his investigation, random motions and collisions might be ignored, also any background interplanetary or terrestrial atmospheric gas.

He found that h, the disturbance field in the free space (p. 35), is $\frac{1}{2}H_{ps}$, where H_{ps} denotes the intensity of the permanent field at the front surface of the stream (cf. eqn. 32 above). The electric field E and the total magnetic field H respectively vary (p. 42), in the stream, approximately as

$$e^{-2(z-z_0)/\lambda}, \qquad e^{-(z-z_0)/\lambda},$$

where

$$\lambda = (m_e c^2/4\pi n_e e^2)^{1/2} = 5.3 \times 10^5/n_e^{1/2}\text{cm}. \tag{3.78}$$

As $n_e(= n_i)$ ranges from 1 to 10000, λ ranges from 5.3 km to 53 meters. Thus the shielding, current-bearing surface layer is very thin compared with the earth's radius a.

The front of the stream comes to rest at a distance z_0, given (p. 35) by

$$z_0 = (3H_0/2)^{1/3}(8\pi n_i m_i W^2)^{-1/6}a, \tag{3.79}$$

where W (taken in numerical illustrations to be 10^8cm/sec) is the initial stream speed. Taking $H_0 = 0.3$ gauss, this gives

$$z_0/a = 8.9 n_i^{-1/6}. \tag{3.80}$$

This is rather insensitive to changes of stream density $n_i(= n_e)$.

The forward motion of the ions (protons) is stopped almost entirely by the electrostatic field, which draws the electrons almost to the same point (p. 33). The motion of an ion-electron *pair* is the same as if both particles had the same mass, $(m_e m_i)^{1/2}$.

The maximum magnetic disturbance h caused by the stream is given by

$$h = \tfrac{1}{3}(8\pi n_i m_i W^2), \tag{3.81}$$

or (for $W = 10^8$cm/sec in a proton-electron gas)

$$h(\text{in } \gamma) = 21.5(n_i)^{1/2}. \tag{3.81a}$$

The following are some values of z_0/a and h for a few values of the stream density

n	1	10	100	1000
z_0/a	8.9	5.9	4.0	2.8
h(in γ)	21	71	215	675

Even the value of h for $n = 100$ is exceptionally high. As an actual solar stream folds somewhat round the earth, it is likely to give a greater disturbing field for the same distance of closest approach (though not so great as for a cylindrical stream, that encloses the permanent field on all sides). The addition to h, in the actual case, by earth currents set up by the rapid change of h, would raise the initial increase of the permanent field (for $n = 100$) to around 400 γ, a value rarely experienced (in low and middle latitude) even during great storms. Thus n seems likely at most to be of order 100/cm³.

The magnetic energy density of the permanent field at the stream face, when this has reached a nearly steady condition, corresponds (§ 2.13) to

$$H_{ps}^2/8\pi = n_i m_i W^2. \tag{3.82}$$

At this stage, therefore, the permanent field magnetic energy density at the stream face equals twice the kinetic energy of the stream. As Martyn (1951) first indicated, such an equation as (82) can also be interpreted in terms of the magnetic pressure of the permanent field, at the stream front. This pressure is absent, or nearly so, behind the thin current-bearing layer of the stream.

The motion of the particles near the stream front is difficult to follow, even in the unreal but simplest case when $m_e = m_i$, so that there is no charge separation and no electric field. The retardation when the masses are unequal exceeds what it would be if both were equal to m_i; this is because of the additional electrostatic retardation when the masses are unequal.

Ferraro (p. 38) gives 0.62 C_{St} as the distance z_1 at which overtaking of the fore by the rear particles begins, where C_{St} refers to mass $(m_i m_e)^{1/2}$—a minor algebraical error in his equation (75), carried forward in his later equations (77–79, 81–84, 138, 139), is here corrected; for atomic hydrogen plasma, with $W = 10^8$cm/sec, this is 700a.

The speed w of advance of the stream front towards its maximum penetration of the permanent field is given (p. 36) by

$$w/W = 1 - (z_0/z)^3,$$

somewhat analogous to (35b) above, for the cylindrical sheet, for $b = 0$. Most of the retardation, and of the growth of the disturbing field h, occur within about two of the time units z_0/W appropriate to this problem. For $n = 1$, $h = 21.5$ γ, this unit is 56 seconds.

The approximations made in the determination of the particle motions begin to fail (p. 45) near the thin foremost layer which is positively charged, but Ferraro inferred that the ions lose about half their initial kinetic energy in this region, and the electrons gain about the same amount of energy in approaching its rear boundary. The other half of the ionic kinetic energy supplies the energy of the magnetic field (p. 45). The lateral motion and energy of the electrons just behind the positively charged layer can greatly exceed their original values, in the ratio $(m_i/2m_e)$.

Ferraro did not calculate the orbits of the particles, or the lateral displacement of their asymptotic lines of motion.

3.10. Instabilities

The study of the impact of solar plasma upon the geomagnetic field is still in a primitive rudimentary state. Something has been learnt about its probable nature and consequences in the highly idealized conditions considered—neglecting the thermal motions of the particles relative to their mean motion, and neglecting any interplanetary magnetic field or any magnetic field transported from the sun by the plasma itself. The importance of such magnetic fields has been stressed repeatedly by Alfvén. The problems without them are so complex that it is natural to ignore them in first studying the plasma motion.

The solution of the cylindrical plasma sheet problem is mathematically pleasing, and it seems to illuminate some features of the sun-earth problem. The "particle" approach to the subject probably still has something worth while to contribute, but a hydromagnetic, continuous-medium approach needs also to be followed.

The idealized problems thus far considered in this field may be likened to the many problems solved early in the development of classical hydrodynamics. They gave some familiarity with the subject, but they need to be extended to include other features of the plasma, including transported magnetic fields and instabilities leading to turbulence.

In recent years many workers have attacked the problem of the form of the hollow formed by the geomagnetic field in the solar plasma. Dr Spreiter, an active worker in this regard, has aided me in preparing the following account of the present state of the subject.

3.11. The steady-state form of the hollow in the stream

The discussion of the impact of a stream of plasma upon the geomagnetic field led Chapman and Ferraro (1931, 1932) to the conception of a steady state of flow of the gas past the earth, around a region of constant form, shielded from the plasma. This hollow in the plasma is bounded by a surface S, and electric currents flow in a thin boundary layer bordering the surface. They shield the rear plasma from the field, and intensify the earth's field throughout most of the hollow. (If the plasma is taken to be of infinite electrical conductivity, and starts from the sun devoid of any magnetic field, no other magnetic field, such as that of the earth, can penetrate it). The steady state is possible for uniform plasma flow, only if (as here) we ignore the daily rotation of the earth dipole about the geographic axis.

Martyn (1951) pictured the steady plasma flow as exerting a constant pressure on the magnetic field, whose own pressure maintains equilibrium. Considering normal incidence on the field, which reverses the uniform approach velocity $\mathbf{v}$ of the plasma particles, the stream exerts a pressure p_0 given by:

$$p_0 = 2nmv^2 \qquad (3.85)$$

Here m denotes the mass of each ion-electron pair, and n the number density of such pairs. Thus Martyn wrote:

$$H^2/8\pi = p_0 \tag{3.86}$$

For illustration, treating the earth's field as that of a dipole (of moment $M = 8.1 \times 10^{25}$), and ignoring the curvature of the dipole field lines and of the hollow surface, Martyn took H to be given by:

$$H = M/r^3\,, \tag{3.87}$$

namely the dipole field intensity in the equatorial plane. From (86) he then inferred the distance at which the stream would be stopped, namely

$$r = (M^2/8\pi p_0)^{1/6} \tag{3.88}$$

For example, if $v = 1000$ km/sec. and the ions are protons, this gives

$$r = 5.00 \times 10^9 n^{-1/6}\text{cm} = 7.87 n^{-1/6}$$

earth radii. This conception of the problem was valuable and illuminating, but the numerical illustration needs correction, because (87) does not include the contribution to **H** made by the surface electric currents. If the plasma surface remained plane, as in Fig. 2, the value of H at the point midway between the dipole and its image would be $2M/r^3$; then (86) adds a factor $2^{1/3}(= 1.26)$ to the right of (88). Further correction, of course, is required because the plasma surface is not plane but curved.

Chapman and Ferraro did not determine the shape of the hollow surface S. After they had accurately solved the cylindrical sheet problem, corresponding to a transient magnetic disturbance, Ferraro (1952), as described in §9, obtained an approximate solution of a model "plane" problem of steady plasma volume flow, into a magnetic field $\mathbf{H}_p$ that varied only along the flow direction. He considered both the transient conditions of approach and the subsequent steady state. The thickness of the surface current layer was found to range from a few meters to a few km; it depends inversely on $n^{1/2}$. The current corresponds to a slight difference between the paths and velocities of the opposite charges in the plasma, as these are deflected by the electric and magnetic fields in the thin surface layer. In the steady state he found that at the surface, **H** has the value $\mathbf{H}_s$ given by:

$$\mathbf{H}_s = 2\mathbf{H}_p \tag{3.89}$$

He considered that the plasma pressure p_0 is balanced half by the transverse pressure of the surface magnetic field, and half by a retarding electric field in the surface layer (caused by the advance of the ions beyond the electrons, involving charge separation in the layer). Thus he gave the equation (his p. 36)

$$H_s^2/8\pi = \tfrac{1}{2}p_0 \tag{3.90}$$

Dungey (1958) showed that the pressure p_0, the change of momentum of the plasma at reflection, must be wholly borne by the magnetic pressure, as in (86), where H denotes the total field and not merely H_p.

Because the current layer is so thin compared with the size of the cavity, some features of the situation can be elucidated by treating the thickness as infinitesimal. Then the magnetic field will be discontinuous at the surface. The shape of the hollow has been studied on this basis by several authors. They formulate the problem as follows:

In the hollow

$$\text{div } \mathbf{H} = 0, \qquad \text{curl } \mathbf{H} = 0, \tag{3.91}$$

and $\mathbf{H}$ is finite and continuous, except for the singularity at the dipole. At any point P on the hollow surface, the plasma particles are supposed to be specularly reflected at the angle of incidence χ; thus the small difference of velocity change, as between ions and electrons, is ignored, although it is an essential feature of the problem, because it supplies the surface current; cf. § 3–6. The normal surface pressure of the plasma at P is therefore $p_0 \cos^2\chi$, and this is balanced by the magnetic pressure $H_s^2/8\pi$; thus:

$$H_s^2/8\pi = p_0 \cos^2\chi \tag{3.92}$$

The equations (91, 92), have been used by Dungey (1958), Zhigulev and Romishevskii (1959 a, b) and Hurley (1961 a, b) to obtain exact solutions (according to this formulation) of certain two-dimensional problems of steady uniform plasma-flow on to various types of permanent magnetic field (§ 12). In these cases it was possible to obtain analytical expressions for $\mathbf{H}_c$, $\mathbf{H}$, $\mathbf{H}_s$ and the form of the surface, where $\mathbf{H}_c$ denotes the field of the surface currents, in the hollow.

The real three-dimensional problem has not been solved exactly, even numerically. But it has been numerically solved approximately, on the basis of (91) and the equation

$$4f^2H^2_{ps}/8\pi = p_0 \cos^2\chi, \tag{3.93}$$

wherein f is taken to be constant. This equation is a valid transformation of (92), but f is a function of position on the surface. Thus, just inside the hollow, at any surface point P,

$$\mathbf{H} = \mathbf{H}_s = \mathbf{H}_{ps} + \mathbf{H}_{cs} \tag{3.94}$$

In general $\mathbf{H}_{ps}$ and $\mathbf{H}_{cs}$ will differ in direction, but in any case we may write

$$f = H_s/2H_{ps}, \tag{3.95}$$

the relation between (92) and (93). Ferraro (1960b) suggested the use of this approximation, taking f to be constant; but in an application he found the result to be "much too crude" (his p. 3953). The use of an incorrect value at any point will contravene the condition that the normal component of $\mathbf{H}$ is zero there.

In the plane problem considered by Ferraro (1952), $\mathbf{H}_{cs} = \mathbf{H}_{ps}$, and therefore $H_s = 2H_{ps}, f = 1$. In the two-dimensional problems described in § 12, $\mathbf{H}_{cs}$ and $\mathbf{H}_{ps}$ are in the same direction, so that

$$H_s = H_{ps} + H_{cs}, \qquad f = \tfrac{1}{2}(1 + H_{cs}/H_{ps}), \tag{3.96}$$

but $H_{cs} < H_{ps}, f < 1$, and f varies over the surface.

The three-dimensional problem has been discussed by Beard (1960) taking $f = 1$, and by Spreiter and Briggs (1962c), using (93); they took alternative values of f (1 and 0.75; the latter is equivalent to taking $\mathbf{H}_{ps} = \frac{1}{2}\mathbf{H}_s = \mathbf{H}_{cs}$). Beard (1960) attempted to estimate the value of f by considering the relation of the surface field $\mathbf{H}_{cs}$ of an adopted distribution of current intensity $\mathbf{i}$, over a hemispherical surface, to the local value of $\mathbf{i}$; the adopted current distribution was not related in any defined way to a plasma flow on to a permanent magnetic field. The two-dimensional problems discussed in § 12 give a better means of illustrating the probable inaccuracy of the assumption $f = 1$. However, Ferraro (1960 a, b), Obayashi and Hakura (1960), Piddington (1960), Dungey (1961), Parker (1961) and Spreiter and Briggs (1962a; see also 1962b) used the equation

$$H_s^2/8\pi = \tfrac{1}{2}p_0 \cos^2\chi,$$

a simple generalization of Ferraro's incorrect boundary equation (90) instead of (92) or (93). This does not affect the approximately deduced *form* of the surface, but makes it too large by a linear scale factor $2^{1/6}(= 1.123)$; cf. Martyn's equation (88). Thus their results require a scale reduction of 11 percent.

3.12. Exact solutions of related problems: two-dimensional cases

Three exact solutions of the two-dimensional form of the problem have been given, in which the earth point-dipole is replaced by a line-dipole. In this case conformal mapping can be applied to the plane of the complex variable $x + iy$, normal to the line dipole. The plasma velocity $\mathbf{v}$ is taken to lie in this plane. Zhigulev and Romishevskii (1959b) determined the form of the boundary (but not the distribution of the magnetic field) in the form of a series expansion; Hurley (1961b) gave closed expressions for the surface and the field. So also did Dungey (1961), whose method, however, unlike that of the other authors cited, was limited to the case of normal incidence of the plasma ($\chi = 0$). For this case Dungey's results agree with Hurley's, though they differ in form. Dr. Spreiter informs me that he found numerical results calculated from the series formula given by Zhigulev and Romishevskii to agree closely with Hurley's results; he was unable to prove their identity analytically.

Fig. 9 (Hurley 1961b) shows the expected shape of the cavity in the line-dipole

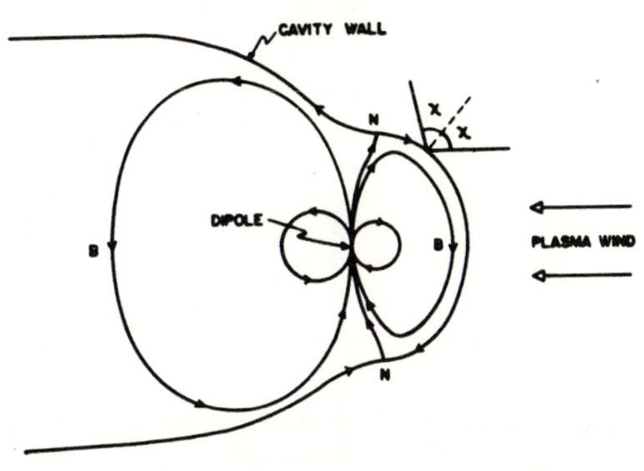

Fig. 3.9. Expected cavity shape and magnetic field configuration. (Hurley, 1961, p. 855.)

case; it shows also typical lines of force in the cavity, and the special line of force from the neutral point N, where $H = 0$; this is a point on a neutral *line* on the cylindrical boundary. It corresponds to the neutral points Q in Fig. 2 for the flat-faced stream impinging on the field of a point dipole.

Hurley's and Dungey's formulae for the boundary curve for normal impact on the field of the line dipole are expressed in terms of the length unit $M^{1/2}(\pi/8p_0)^{1/4}$, where M denotes the dipole moment per unit length. The origin is taken at a point on the dipole. The equation of the upper part of the curve ($y > 0$) as given by Hurley is:

$$\pi x = 2 + \ln \frac{(1-u)^u}{(1+u)^u} \frac{(1+v)^2}{v^2} \tag{3.97}$$

Here $u = \sin\phi$, $v = \cos\phi$ and $u = y$ for the part of the curve between the upper and lower neutral points. For the remaining upper part of the curve, $u = 2 - y$. The vertex of the cavity, on the line $y = 0$, corresponds to

$$\phi = 0, \qquad x = 2(1 + \ln 2)/\pi = 1.07.$$

The neutral point N corresponds to

$$\phi = 90°, \qquad x = 2(1 - \ln 2)/\pi = 0.197, \qquad y = 1.$$

Thus the part of the boundary between the two neutral lines does not differ much from half a circular cylinder. The curve marked Exact in Fig. 10 (Hurley 1961b) shows the boundary as drawn from the above formula by Hurley; Dungey's formulae are equivalent (except that the x axis is taken in the opposite direction), but in his illustration of the boundary the scales of x and y differ, owing to a numerical oversight. In Fig. 10 the plasma flow is from the right. The neutral line lies on the day side of the dipole.

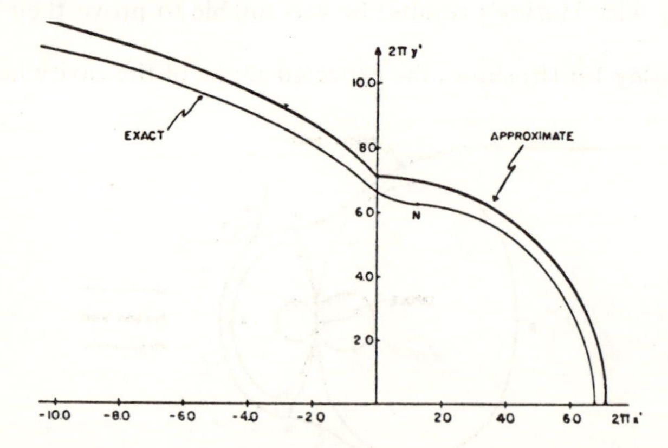

Fig. 3.10. Exact cavity surface as determined by Hurley; and Beard's approximation to it; two-dimensional case.
(Hurley, 1961, p. 857.)

Hurley (1961b, p. 857) found the distance of the vertex V of the cavity from the ine dipole, exactly calculated, to be

$$1.35[2M/(8\pi p_0)^{1/2}]^{1/2}.$$

\s $H = 2M/y^2$ at distance y along Oy, this value of y corresponds, according to 93), (wherein here $\chi = 0$) to $f = 1.35^2/2$ or 0.91. If f is taken to be 1, the distance ?V is over-estimated by four per cent.

Dungey discussed the form of the line of force from the dipole to the neutral oint. Its equation may be expressed thus:

$$y = 2\phi/\pi,\ x = (1 - \operatorname{cosec}\phi)\ln(1 - \sin\phi) + (1 + \operatorname{cosec}\phi)\ln(1 + \sin\phi) - 2 \quad (3.98)$$

Near the origin of the dipole this equation approximates to $x = \pi y^2/12$. Dungey compared this with the line of force in the combined field of the line dipole and an mage line dipole at distance $2R$. The line of force through the neutral point has the quation $x^2 + y^2 = 2Rx$ (relative to an origin at the real line dipole). Near the origin his approximates to $y^2 = 2Rx$. Taking R to have the same value as the distance to he exactly calculated vertex of the cavity, namely $R = 1.07$, this value of y^2/x is 2.14, s compared with $12/\pi$ or 3.83 in the exact case. Thus the flat-faced model of the ield of the line dipole and the surface currents (analogous to Fig. 2, except that that efers to a point dipole and its image) gives for y^2/x little more than half the true value. This means that the latitude at which the critical line of force cuts the earth, of radius a, is under-estimated by the approximate calculation. Writing $y^2/x = k$ or the critical line of force near the origin, the point at which it cuts the earth, when a/R is small, has the x coordinate $\frac{1}{2}a^2/k$, and the colatitude arc sin $\frac{1}{2}a/k$. For example, considering a plasma flow strong enough to bring the vertex of the cavity to 10 earth-radii from the dipole, so that $a = 1.07/10$ in the length units here used, the exact theory ($k = 3.83$) gives the colatitude 0.9°, as against the approximate value for $k = 2.14$) of 1.4°. These latitudes are very high; they depend on the intensity of the plasma stream, which governs the length unit; the greater the value of p_0, he smaller the unit and the larger the value of the earth's radius in this unit; hence he lower the latitude of entry of the critical line of force.

Dungey (1958) considered the latitude, using the flat-faced approximation, or a *point* dipole (loc. cit. p. 147). In this case dH, the field of the image dipole, at the earth, is $M/8R^3$; taking it for example to be 200γ, R/a is 5.7; using an approximate formula for the critical line of force, he found the latitude of entry to be 72° n this case (the cosine of the latitude varies as $dH^{1/6}$). The above comparison suggests hat the true latitude for the curved-faced plasma would be greater.

These points of entry relate to the "horns" mentioned by Chapman and Ferraro n their discussion of Fig. 2. The significance of the horns for geomagnetic disturbance has been considered by various writers, in connection with studies of polar disturbance. Among these may be named Mayaud (1955), and more recently Lebeau and Schlich (1962). The horns may be channels along which particles and also hydromagnetic waves may travel into the earth's atmosphere. But it seems unlikely that the horns are important for the aurora, which requires line sources instead of point sources.

Hurley gave a general expression for the field intensity at any point in the cavity formed by the line dipole of Fig. 10; he showed that at the dipole the field of the surface currents in the plasma is $7(\pi^3 p_0/8)^{1/2}/9$. For this to be 20γ, p_0 must be 1.7×10^{-8}dyne/cm^2; for a proton-electron stream with a speed 10^8cm/sec, the necessary value of n is approximately 1.

Hurley also gave diagrams showing the form of the cylindrical boundary of the cavity for plasma flow when the dipole axis is normal to the direction of v or $-v$, and when it is tilted by 37° from this position. The terrestrial point dipole is tilted at almost this angle at a certain Greenwich time at each solstice.

Another two-dimensional problem of plasma flow towards a two-dimensional magnetic field distribution has been considered approximately by Ferraro (1960), and solved exactly by Zhigulev (1959a) and Hurley (1961a). The field is that of an infinitely long line current. Comparing the approximate solution found by using (93), with the exact solution, the agreement is found to be poor. The equation of the boundary curve is given by Zhigulev as:

$$y/a^{1/2} = \text{arc cos}\,(\tfrac{1}{2}e^{-x/a^{1/2}} - 1). \tag{3.99}$$

Hurley gave it in the parametric form

$$x = -\,2a^{1/2}\ln\frac{(u^2 + v^2)^{1/2}}{2u}, \qquad y = -\,2a^{1/2}\,\text{arc tan}\,(v/u) \tag{3.100}$$

He also gave a diagram of the cavity boundary for this problem; here there is no neutral point.

In these problems the cavity is open on the side opposite that of impact. The same is true for the cavity considered by Chapman and Ferraro for the earth dipole. When a projectile is shot into a fluid with great speed, a cavity is formed behind it, but the cavity is closed if the penetration is sufficient. Relative to the plasma from the sun, the earth enters it with great speed, and a cavity is formed, not by the solid body of the earth but by the magnetic field that surrounds it, and shields it from the plasma. Here too it may be expected that the cavity will close, and that the length on the side of the earth away from the sun will depend at least partly on the random speeds (or the temperature) of the plasma. This provides a lateral pressure on the magnetic field behind the earth, and may be expected to close the cavity where the pressure exceeds that of the field. As yet we do not know the temperature of the plasma or the length of the cavity behind the earth.

3.13. Somewhat analogous axially symmetrical three-dimensional problems

Midgley and Davis (1962) and Slutz (1962) have independently discussed and numerically solved an analogous problem of equilibrium between a plasma and a magnetic field. The field is that of a point dipole, but the plasma is considered not as *flowing* on to the field, but as static and exerting an isotropic pressure p'_0 on the field. The cavity in this case has rotational symmetry about the dipole axis. The equations used are (91) and (100), with

$$H_s^2/8\pi = p'_0.$$

Different methods of numerical solution were used, and the results agree within 1 per cent on the dipole axis, and within 3 per cent in its equatorial plane: the differences are within the expected limits of accuracy of the computation. Fig. 11

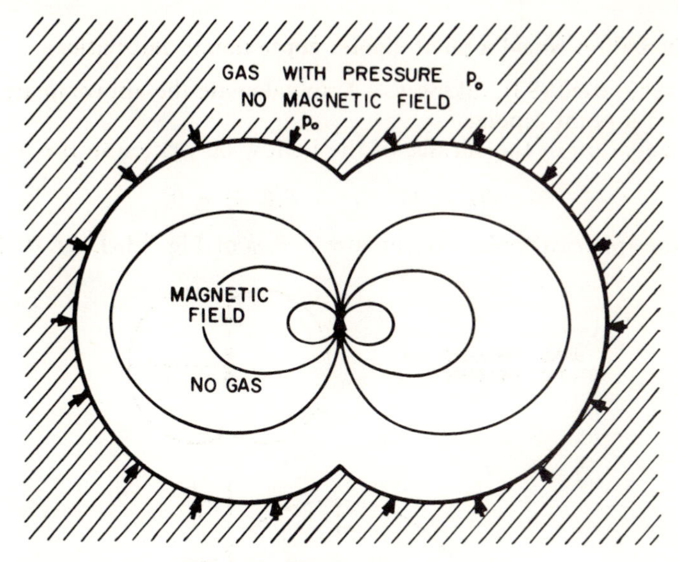

Fig. 3.11. (Slutz, 1962a, p. 507.)

(Slutz 1962) shows the form of the cavity, in terms of a parameter r_0 which represents the equatorial distance at which the undisturbed dipole field would balance p'_0; thus $r_0 = M^{1/3}(8\pi p'_0)^{1/6}$. The surface of the cavity in the equatorial plane does not approach the earth to this distance: the distance is $1.37 r_0$. This is because of the field of the currents that flow in the surface, and support the plasma by magnetic pressure.

The solution cannot be exact near the cusps on the dipole axis, because H there must be zero, contrary to the equilibrium equation assumed. "The behavior at these points, which correspond to the neutral points in the problems already described, must differ from that corresponding to the equation (92). Such an indentation may well occur at these neutral points if the plasma is supposed to have some gas pressure due to thermal motions, as well as its mean motion. Then a pressure term would have to be added on the right of (92)." (Spreiter).

Slutz has extended his work described above by calculations in which (92) is replaced by

$$\begin{aligned} H_s^2/8\pi &= 2nmv^2\cos^2\chi + p'_0, \qquad \text{for } \cos\chi \geqslant 0, \\ &= p'_0 \text{ for } \cos\chi = 0. \end{aligned} \tag{3.101}$$

He took $p'_0/2nmv^2$ to be of the order about 1 percent, and thus obtained a closed cavity with two neutral points on its surface; naturally the cavity extends further from the earth on the side away from the sun than on the sunward side; see Fig. 12 (Slutz 1962b).

These numerical calculations now seem to promise a complete solution of the form of the cavity in the steady state; they do not yet determine the magnetic field in the cavity. For this it is still necessary to resort to approximate solutions.

3.14. The hollow round a point dipole: approximate calculations

Beard (1960) used (93), taking $f = 1$, to calculate the approximate form of the hollow around a point dipole.

Let the equation of the surface of the hollow be written in the forms

$$f(r, \theta, \phi) = r - F(\theta, \phi) = 0 \tag{3.102}$$

in terms of the spherical polar coordinates r, θ, ϕ of Fig. 13. Let $\mathbf{r}$, $\boldsymbol{\theta}$, $\boldsymbol{\phi}$ denote the

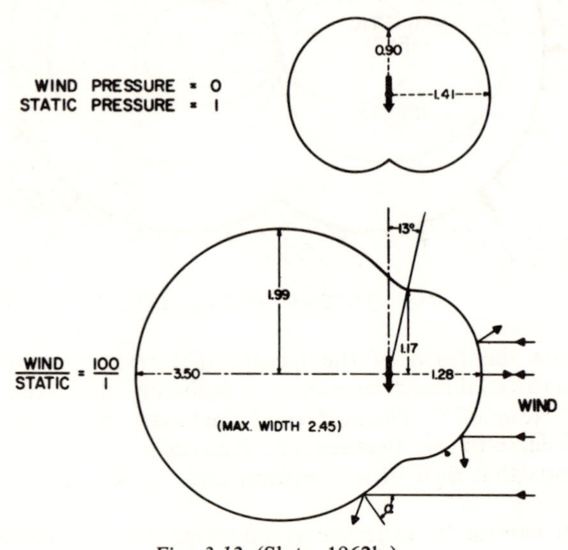

Fig. 3.12. (Slutz, 1962b.)

unit vectors at P in the directions of increasing r, θ, ϕ respectively. Any displacement $\mathbf{ds}$ from P can be expressed thus:

$$\mathbf{ds} = \mathbf{r}\, dr + \boldsymbol{\theta} r\, d\theta + \boldsymbol{\phi} r \sin\theta\, d\phi \tag{3.103}$$

If this displacement is along the surface (102),

$$f_r\, dr + f_\theta\, d\theta + f_\phi\, d\phi = 0 \tag{3.104}$$

where

$$f_r \equiv df/dr, f_\theta \equiv df/d\theta, f_\varphi \equiv df/d\phi$$

The unit vector $\boldsymbol{\nu}$ normal to the surface is perpendicular to all such displacements: hence if

$$\boldsymbol{\nu} = l\mathbf{r} + m\boldsymbol{\theta} + n\boldsymbol{\varphi}, \tag{3.105}$$

the condition of normality, $\boldsymbol{\nu} \cdot \mathbf{ds} = 0$, gives

$$l\,dr + mr\,d\theta + nr\sin\theta\,d\phi = 0 \qquad (3.106)$$

As (104, 106) are valid for an infinity of ratios $dr{:}d\theta{:}d\phi$, it follows that

$$l = af_r, \qquad m = af_\theta/r, \qquad n = af_\phi/r\sin\theta \qquad (3.107)$$

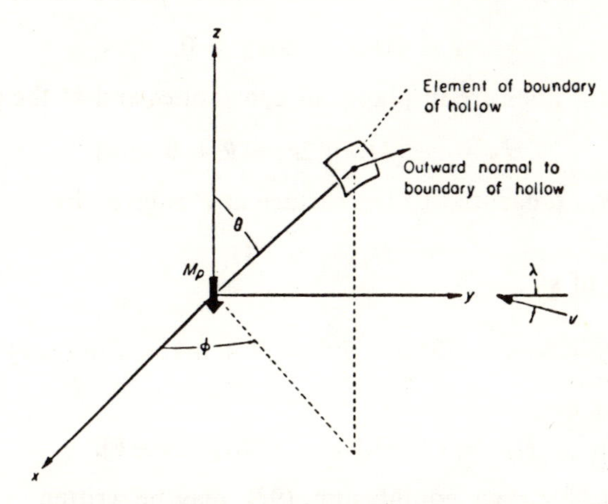

Fig. 3.13. View of coordinate system. (Spreiter and Briggs, 1962a, p. 38.)

where

$$a^2(f_r^2 + f_\theta^2/r^2 + f_\phi^2/r^2\sin^2\theta) = 1.$$

Note that

$$f_r = 1, \qquad f_\theta = -r_\theta, \qquad f_\phi = -r_\phi, \qquad (3.108)$$

where

$$r_\theta = \partial F(\theta, \phi)/\partial\theta, \qquad r_\phi = \partial F(\theta, \phi)/\partial\phi. \qquad (3.109)$$

Because $l^2 + m^2 + n^2 = 1$,

$$a^{-2} = 1 + (r_\theta/r)^2 + (r_\phi/r\sin\theta)^2. \qquad (3.110)$$

In Fig. 13 the direction of $\mathbf{v}$, the plasma velocity, lies in the plane $\varphi = 270°$. Hence

$$\mathbf{v} = -\mathbf{j}v\cos\lambda + \mathbf{k}v\sin\lambda$$

and

$$\mathbf{j} = \mathbf{r}\sin\theta\sin\phi + \boldsymbol{\theta}\cos\theta\sin\phi + \boldsymbol{\phi}\cos\phi, \qquad \mathbf{k} = \mathbf{r}\cos\theta - \boldsymbol{\theta}\sin\theta$$

Hence (Beard 1960) the unit vector of $\mathbf{v}$ $(= \mathbf{v}/v)$ is given by

$$\mathbf{r}(\cos\theta\sin\lambda - \sin\theta\sin\phi\cos\lambda) - \boldsymbol{\theta}(\cos\theta\sin\phi\cos\lambda + \sin\theta\sin\lambda) - \boldsymbol{\phi}\cos\phi\cos\lambda.$$

Thus the angle of incidence χ between $-\mathbf{v}$ and the outward normal $\boldsymbol{\nu}$ to the surface at P, is such that

$$\cos\chi = -\mathbf{v}\cdot\boldsymbol{\nu} = a[\sin\theta\sin\phi\cos\lambda - \cos\theta\sin\lambda$$
$$-(r_\theta/r)(\cos\theta\sin\phi\cos\lambda + \sin\theta\sin\lambda) - (r_\phi/r\sin\theta)\cos\phi\cos\lambda]. \qquad (3.111)$$

The plasma impinges on the hollow surface only at points where

$$\chi \leqslant 90^\circ, \qquad \cos\chi \geqslant 0.$$

The dipole field lines are in planes of constant ϕ, and at the point P

$$H_p = -(M/r^3)(2\mathbf{r}\cos\theta + \boldsymbol{\theta}\sin\theta)$$

The component H_{ps} tangential to the surface at P is given by

$$H_{ps} = \boldsymbol{\nu}'\cdot\mathbf{H},$$

where

$$\boldsymbol{\nu}' = b(m\mathbf{r} - l\boldsymbol{\theta}), \qquad b^{-2} = l^2 + m^2 = 1 + (r_\theta/r)^2.$$

Hence

$$H_{ps} = (M/r^3)[\sin\theta + 2(r_\theta/r)\cos\theta]. \qquad (3.112)$$

The equation of surface equilibrium, (93), may be written

$$H_{ps} = (2\pi p_0/f^2)^{1/2}\cos\chi \qquad (3.113)$$

or, using the above expressions for H_{ps} and $\cos\chi$,

$$\sin\theta + 2(R_\theta/R)\cos\theta = R^3(\cos\chi)/b, \qquad (3.114)$$

where

$$R = r/r_0, \qquad r_0{}^3 = fM/(4\pi nmv^2)^{1/2} \qquad (3.115)$$

Thus r_0 is used as a length unit appropriate to this problem. Its significance is readily seen: for normal impact on the direction of the dipole axis ($\lambda = 0$), the equatorial plane ($\theta = 90^\circ$) and the noon and midnight meridian planes ($\phi = 90^\circ$) are clearly planes of symmetry of the surface; also $r_\theta = 0$ in the plane $\theta = 90^\circ$, and $r_\phi = 0$ in the plane $\phi = \pm 90^\circ$. On the sun-earth line ($\theta = 90^\circ$, $\phi = 90^\circ$), $a = b = 1$, $\chi = 0$, and (115) gives

$$R = 1, \qquad r = r_0. \qquad (3.116)$$

Thus r_0 is the distance OV from the earth's center O to the vertex V of the hollow.

3.14a. The hollow at the equinoxes ($\lambda = 0$)

When $\lambda = 0$, (114) has the special form

$$\sin\theta + 2(R_\theta/R)\cos\theta = (aR^3/b)[\sin\theta\sin\phi - (R_\theta/R)\cos\theta\sin\phi$$
$$-(R_\phi/R\sin\theta)\cos\phi)]. \qquad (3.117)$$

By taking $\varphi = 90°$ we obtain the equation for the noon meridian section of the surface, which may be denoted by s(noon, $\lambda = 0$); its equation is

$$\sin\theta + 2(R_\theta/R)\cos\theta = R^3[\sin\theta - (R_\theta/R)\cos\theta], \tag{3.118}$$

because in this plane $R_\phi = 0$, $a = b$, (or rather $a = \pm b$, on account of the ambiguity of sign of the square roots involved in a and b). The sign must be taken as here, because at the vertex V, $\chi = 0$, $\cos\chi = +1$. The same sign must be valid, by continuity, elsewhere on the surface, at least up to points or lines where $\cos\chi = \mathbf{0}$. The equation (118) for s(noon, $\lambda = 0$) may also be written (Beard 1960)

$$\tan\theta\, d\theta = \frac{2 + R^3}{R^3 - 1}\frac{dr}{R}. \tag{3.119}$$

For the section s(midnight, $\lambda = 0$) by the midnight meridian plane $\phi = -90°$, the signs of a and b in (111, 112) must again be positive, in order that H_{ps} and $\cos\chi$ shall be positive. Then (117) gives the equation of this section as

$$\sin\theta + 2(R_\theta/R)\cos\theta = R^3[-\sin\theta + (R_\theta/R)\cos\theta], \tag{3.120}$$

which is equivalent to

$$\tan\theta = \frac{R^3 - 2}{1 + R^3}\frac{dR}{Rd\theta}. \tag{3.121}$$

The equations (119, 121) have the respective solutions

$$\cos\theta = KR^2/(R^3 - 1), \qquad \cos\theta = KR^2/(R^3 + 1), \tag{3.122a, b}$$

or

$$z = R\cos\theta = K/(1 - R^{-3}), \qquad z = R\cos\theta = K/(1 + R^{-3}). \tag{3.123a, b}$$

Here K denotes an arbitrary integration constant. These two families of curves are illustrated in Fig. 14 (Spreiter and Briggs 1962a). As Beard (1960) concluded, the only appropriate member of the family (123a) is the circular arc ($K = 0$)

$$R = 1 \tag{3.124}$$

The curves (123a) for all other values of K reach the sun-earth line only at either the origin or at infinity. In the midnight meridian plane $\chi > 90°$; that is, the section is not met by the plasma flow. Hence, this part of the curve is irrelevant to our problem. In that plane the section is one of the curves (123b).

It would seem appropriate to choose K for the latter curve so as to join the curve (124) on the dipole axis $\theta = 0$; this would correspond to $R = 1$, $z = 1$, $K = 2$; however, as is clear from Fig. 14, this curve (123b) is unsuitable, because everywhere on it (in the midnight half plane) $\chi > 90°$. Spreiter and Briggs therefore choose K so that the curve (123b) extends into the noon half plane to meet (124). There is only one such curve (123b) that meets (124) and also extends to $y = \infty$ in the midnight plane. This is the curve with the double point on the dipole axis (Fig. 14). As (121) shows, $dR/d\theta = 0$ at $\theta = 0$ for all curves (123), except for the curve for which on the polar axis $R = 2^{1/3}$; (123b) shows that for this curve

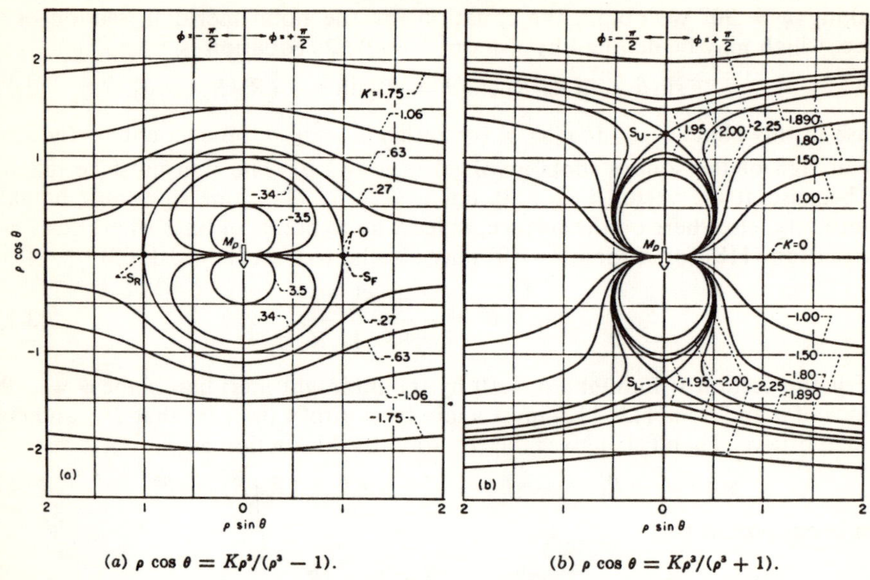

(*a*) $\rho \cos \theta = K\rho^3/(\rho^3 - 1)$. (*b*) $\rho \cos \theta = K\rho^3/(\rho^3 + 1)$.

Fig. 3.14. Integral curves defined by equations 122. (Spreiter and Briggs, 1962a, p. 41.)

$K = 3/2^{2/3} = 1.890$. This curve meets (124) at $\cos \theta = \frac{1}{2}K = 3/2^{5/3}$, $\theta = 19.1°$. Fig. 15 (Spreiter and Briggs 1962a) shows the resulting section of the hollow; it resembles Hurley's section for the two-dimensional case, except for the discontinuity of direction where the curves (123b, 124) meet.. It is not clear, however, that the part of the curve (123b) that lies in the noon meridian half-plane satisfies the postulated condition (93), and to this extent the validity of the procedure is doubtful. The discontinuous change of direction at the point where the two curves join is a feature not shown by the neutral point in Hurley's two-dimensional curve. The uncertainty as to the value of f and its changes, especially in the region of the neutral point, affects the reliability of the positions of this point as shown in Fig. 15.

Beard (1960, 1962) treated the shape of the section by the plane $x = 0$ in a different way, hence his 1960 result differs from that of Spreiter and Briggs (1962b), who illustrate the difference by Fig. 16. Their curve agrees with Beard's, as shown in Fig. 15, in part of the noon meridian half-plane; in the midnight meridian half plane Beard took the section for $R > 2^{1/3} = 1.26$ to be given by (123b) for $K = 1.8$; Fig. 14a shows that if continued for $R < 1.26$ this curve would approach the dipole. Beard (1960) continued the midnight meridian section from the point $R = 1.26$ by the curve $y = \frac{1}{2}(1 + R^{-3})$, a solution of (121) with one side reversed in sign; no reason was given for this choice, but the midnight meridian section was stated to be very approximate, in that the constant of integration in (123b) is determined by matching the daylight solutions "across the region of zero surface current" (namely near the neutral point) "for which the solution has not yet been found precisely". Later Beard remarked (1962, p. 482) that in his first paper "the solution at the poles and onto the dark side of the earth was recognized to be unattainable from equating the particle pressure to the magnetic field pressure if the non-local

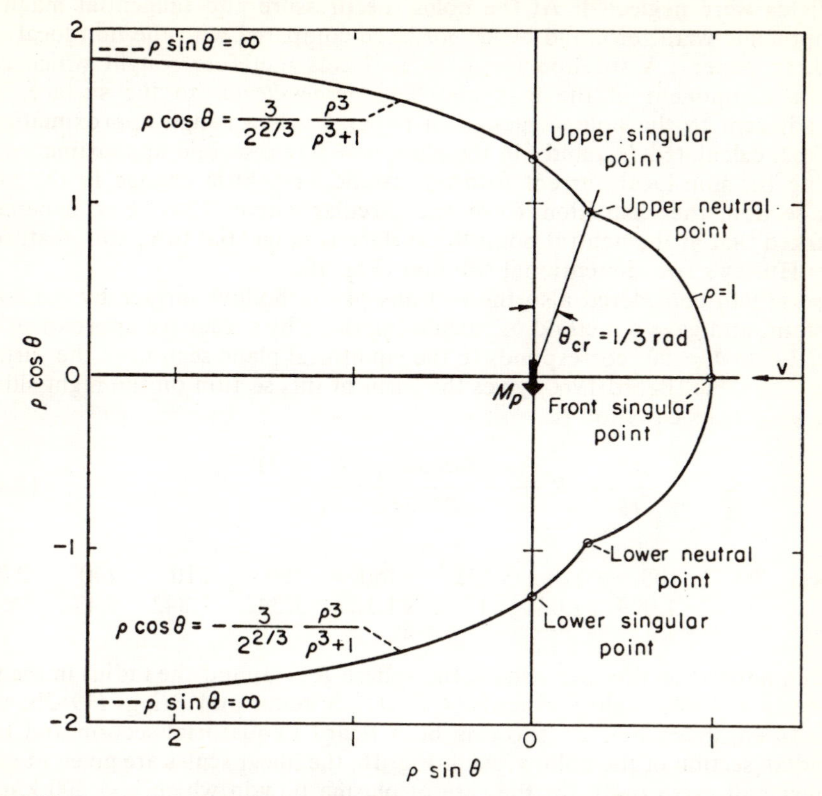

Fig. 3.15. Form of the boundary of the hollow in the meridian plane containing the dipole axis and the sun–earth line; $\lambda = 0$. (Spreiter and Briggs, 1962a, p. 43.)

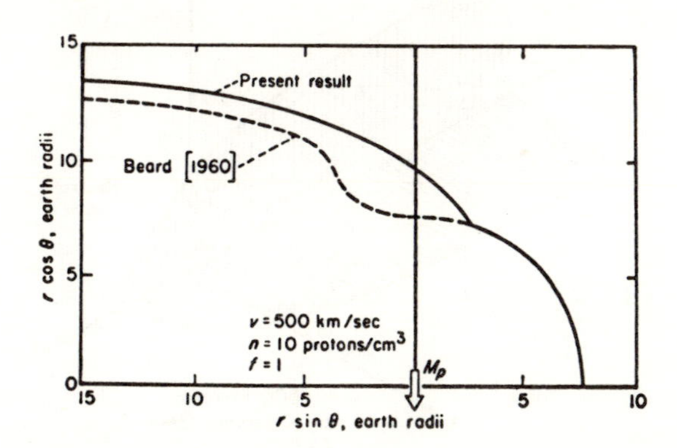

Fig. 3.16. Comparison of present results and those given by Beard (1960) for the trace of the boundary of the hollow in the meridian plane containing the dipole axis and the sun–earth line; $\lambda = 0$, $v = 500$ km/sec, $n = 10$ protons/cm^3, and $f = 1$. (Spreiter and Briggs, 1962b, p. 2984.)

current fields were neglected. At the poles the pressure and tangential magnetic field components were zero, and hence *not* large compared with the non-local current field. In paper 1 a solution for $\varphi = 3\pi/2$ was therefore sought which considered the component of the magnetic field *perpendicular* to the surface. The solution adjacent to the pole suggested in paper 1 was a rough approximation." Beard (1962) calculated H_{ps} and f in the plane $x = 0$ to a second approximation by considering the non-local current field. He found very little change in the noon meridian section, the distortion from the circular curve (124) being 3 percent. He remarked that at the neutral point the surface is tangential to $\mathbf{v}$; this feature is shown by Hurley's two-dimensional solution (Fig. 10).

Beard (1960) considered also the sections of the hollow surface by the cones θ = constant, and gave a method of calculating them by successive approximation. The special case $\theta = 90°$ corresponds to the equatorial plane section of the surface. The following table (Beard 1960) gives the form of this section on the night side of the plane $y = 0$; its equation is

$$\frac{dR}{d\phi} = R\frac{R^6 \sin\phi \cos\phi + (R^6 - 1)^{1/2}}{R^6 \cos\phi - 1} \tag{3.125}$$

φ (degrees)	90	105	120	135	150	180	210	240	270
R	1	1.009	1.031	1.068	1.126	1.342	1.842	3.472	∞

Thus the equatorial section lies outside the sphere $R = 1$, and the radius in the twilight plane ($\phi = 180^0$) slightly exceeds (4/3) OV. Spreiter and Briggs (1962b) used this table to construct Fig. 17, showing both Beard's equatorial section and their own meridian section of the hollow. As in Fig. 16, the linear scales are given numerically in terms of earth radii, for the case of plasma flow in which v = 500 km/sec and $n = 10/\text{cm}^3$.

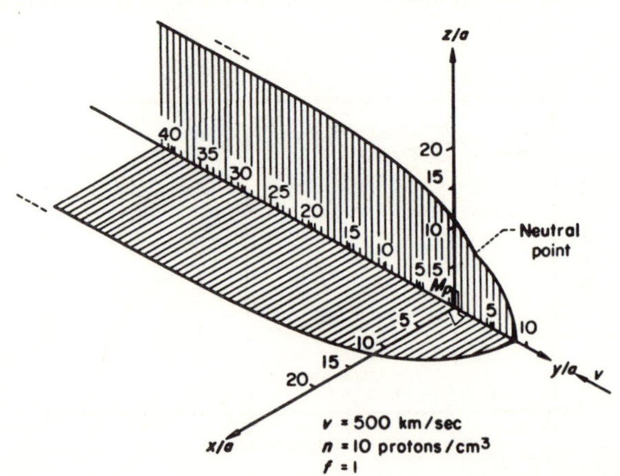

Fig. 3.17. Boundary of the hollow in the equatorial plane and the meridian plane containing the dipole axis and the sun–earth line; $\lambda = 0$. (Spreiter and Briggs, 1962b, 2985.)

3.14b. Oblique incidence of the solar plasma

Beard (1960) obtained the equations (111) and (113) which enable (92) to be expressed for any angle λ of incidence of the solar plasma on the dipole axis. He remarked (his p. 3566) that he had not examined this general case in careful detail, but the principal effect appears to be that the axis of the current sheath is tilted through an angle Φ_h. The nearest distance of the surface to the earth is increased to roughly $(\cos \Phi)^{-1/3} r_0$. "The surface is very little changed otherwise from the $\lambda = 0$ case".

When $\lambda = 0$ and $\phi = 90°$ or $\phi = -90°$, (93) takes the respective special forms

$$\sin\theta + 2(R_\theta/R)\cos\theta = R^3[\sin(\theta - \lambda) - (R_\theta/R)\cos(\theta - \lambda)] \quad (\phi = 90°) \qquad (3.126a)$$

$$\sin\theta + 2(R_\theta/R)\cos\theta = -R^3[\sin(\theta + \lambda) + (R_\theta/R)\cos(\theta + \lambda)] \quad (\phi = 90°) \qquad (3.126b)$$

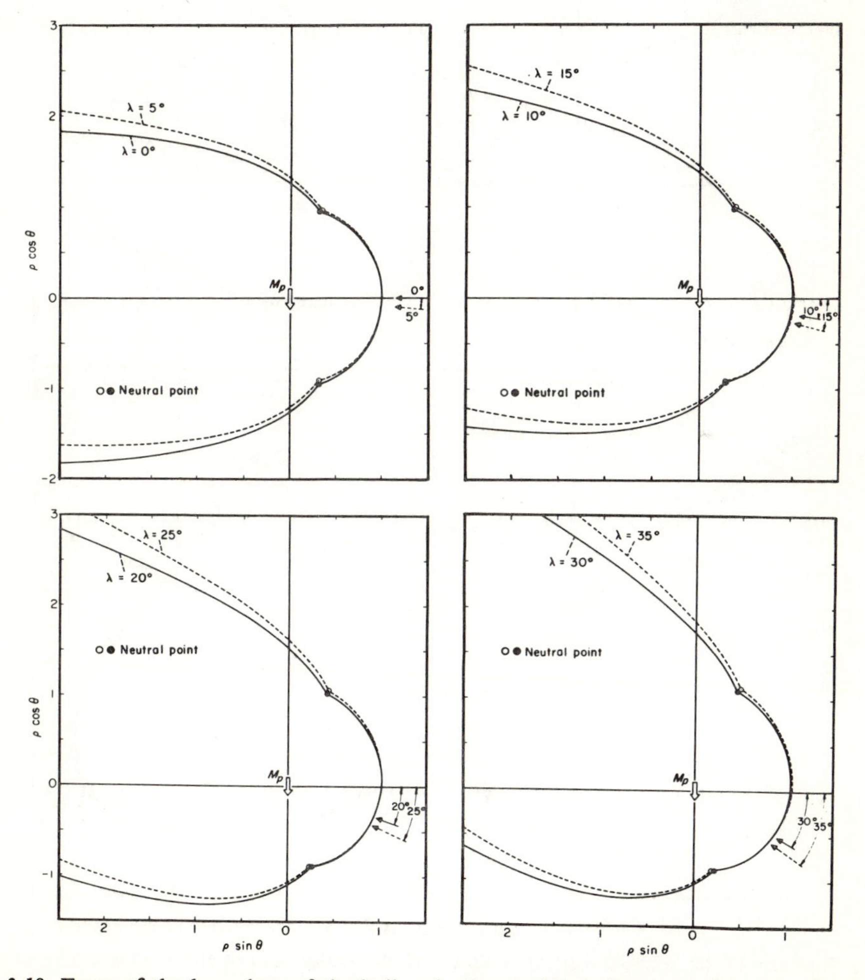

Fig. 3.18. Form of the boundary of the hollow in the meridian plane containing the dipole axis and the sun–earth line; $0° \leqslant \lambda \leqslant 35°$. (Spreiter and Briggs, 1962a, p. 45.)

The corresponding solutions are respectively

$$R^3 \cos(\theta - \lambda) - \cos\theta = KR^2, \qquad R^3 \cos(\theta + \lambda) + \cos\theta = KR^2; \tag{3.127a, b}$$

they reduce to (122a, b) when $\lambda = 0$.

The equations (126) and their solutions have singular points, and by a procedure similar to that followed for the case $\lambda = 0$, and subject to the same question, Spreiter and Briggs (1962a) fitted to (127a) an upper and a lower curve each passing through a singular point of the family (127b). Fig. 18 shows the result thus obtained,

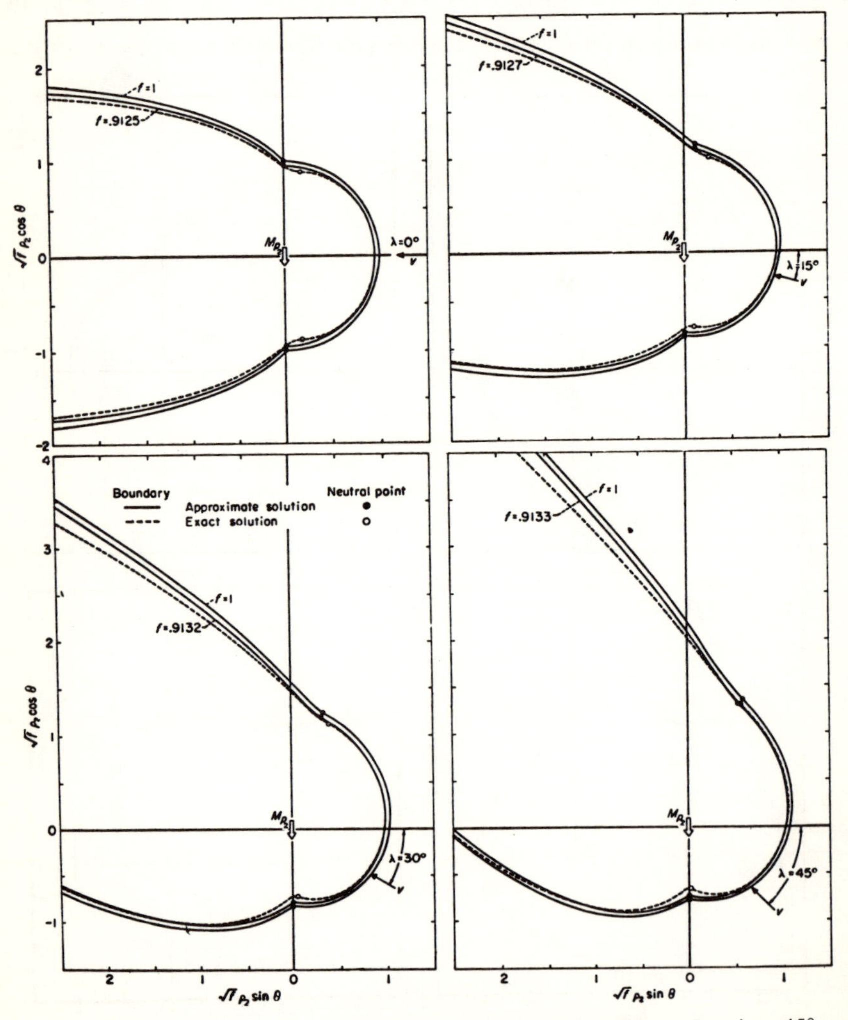

Fig. 3.19. Form of the boundary of the hollow, two-dimensional problem; $0 \leqslant \lambda \leqslant 45°$. (Spreiter and Briggs, 1962a, p. 48.)

for $\lambda = 0$ and for five other values of λ, up to $\lambda = 35°$; the latter approximates to the maximum possible real value.

Spreiter and Briggs (1962a) also applied the equation (93) to the line-dipole (§ 12); when $\lambda = 0$ the equation (124) is valid also for this two-dimensional problem, in the noon meridian half-plane; in the midnight one ($\phi = 90°$), (123b) is replaced by $z = K/(1 + R^{-2})$. In this case $K = 2$ and the two curves join on the dipole axis, disagreeing with the exact solution. Fig. 19 compares the exact and approximate sections by the plane $x = 0$ in this two-dimensional case, for various values of λ; each diagram shows two approximate curves, for $f = 1$ and for $f = 0.913$; the latter values gives better overall agreement with the true section. The curves $f = 1$ give a linear scale about 5 per cent too large. The approximate solutions give the neutral points in slightly erroneous positions, at which the section has an incorrect cusp form; but the general agreement of the approximate and true solutions is decidedly good.

Finally it may be noted that the current in the surface obtained by means of the approximate equation (93) will not fully shield the plasma from the permanent field. Midgley and Davis (1962, p. 503) state that their "moment technique" method of calculation of the surface "gives a net field outside which is about 0.001 of that given by the surface derived using Beard's boundary condition".

4. Magnetic Disturbance and Aurora

This chapter briefly reviews some main features of magnetic disturbances, which when specially intense are called magnetic storms. Some of the outstanding properties of auroras, including their association with magnetic storms, are also described.

4.1. The parts M, S, L, D, of the geomagnetic field

The earth's magnetic field averaged over a year or more is called the main field, and denoted by M. It has a slow secular variation, which need not be considered here. There are also transient variations of the field, solar daily (S), lunar daily (L) and irregular or disturbance (D). These correspond to certain changing magnetic fields, also denoted by S, L and D, superposed on M. Unlike the main field, which proceeds almost wholly from within the earth, the S, L and D fields originate above the earth's surface. Systems of electric current, also denoted by S, L and D, are there generated in different ways and in different regions. The magnetic fields of these currents form the main parts of the S, L and D fields. These parts may be denoted by S_e, L_e and D_e, the subscript e signifying *external*. These changing S_e, L_e and D_e fields induce electric currents inside the earth. These current systems, S_i, L_i and D_i, where the subscript i denotes *internal*, have magnetic fields also denoted by the same symbols S_i, etc. These make up the remainder of the S, L and D fields; thus $S = S_e + S_i$ and so on. By means of spherical harmonic analysis it is possible to determine the external and internal parts of the S, L and D fields at the earth's surface. The ratio of the internal secondary part to the primary external part depends on the rapidity of the changes; the two are more nearly equal, the more rapid the changes. The general effect at the earth's surface is that the internal currents increase the horizontal component of the field considered (S, L or D), and they reduce the vertical component.

More than a hundred magnetic observatories continuously record the changes in the total field, in three magnetic elements. These are usually H the horizontal component, its declination d^0 from geographic north, and the vertical force Z (positive if downward). Sometimes the records refer to Z and to X, Y, the north and east horizontal components ($X = H \cos d$, $Y = H \sin d$). (In §1.3 the declination is denoted by D, but in this chapter, where disturbance is the special subject of discussion, it is more convenient to use D for disturbance and d° for declination).

Hourly mean values are measured from the records, and are set out in monthly tables. Each row refers to one day, and there are 24 hourly columns. Means of the hourly columns are taken for all days of the month, and also for five quiet (q) days, and for five disturbed (d) days. The q and d days are chosen internationally (§1.9), on the basis of the international daily magnetic character figures C or Cp, for each Greenwich day. From each set of means the average value for each set is subtracted. This gives in each case a sequence of 24 values, whose sum is zero; the sequence is called a mean daily inequality, and the three sequences obtained from all days, q days and d days are denoted by S_a, S_q and S_d respectively. Such sequences are obtained for each element, and the element to which any such sequence refers can be indicated as in $S_a(H)$ or $S_q(X)$. The totality of all such sequences from all stations represents our observed data for the study of the solar daily variation S in that month, from the three different kinds of day.

From these data for the two horizontal components H and d^0 or X and Y it is possible to compute the magnetic potential of the corresponding field S_a, S_q or S_d at the earth's surface. This potential can be expressed as a series of spherical surface harmonics; so also can the distribution of the scalar function $S(Z)$, for a, q or d days. By combination of the two series of spherical surface harmonics it is possible to obtain the potentials, at the earth's surface, of the external and internal parts of the said field, e.g., S_{ae} and S_{ai}, and similarly for the L and D fields.

The fields S_a, S_q and S_d differ in the amount of disturbance that affects the corresponding sets of days. There may be some disturbance even on the q days, especially in sunspot maximum years. The solar daily variation on days when there is no disturbance at all is denoted by Sq. In low and middle latitudes, particularly at or near sunspot minimum, the S_q variation and field well represent Sq. In auroral latitudes (above 60° latitude) S_q is usually affected by disturbance even on most q days; there Sq can be found only from the S sequences on rare exceptionally quiet days.

The difference-sequences symbolized by $S_d - Sq$, $S_a - Sq$ represent the contribution of the D field to the solar daily variation, in the average of d days or of all days; naturally it is in general greater in the sequence S_d–Sq than in S_a–Sq. In low and middle latitudes Sq is taken to be given by S_q in forming these difference sequences. The contribution of the D field to the solar daily variation and field is denoted by SD; as represented by the difference sequences S_d–S_q and S_a–S_q, it may be denoted by SD_d or SD_a. The higher the latitude, up to the latitudes of the auroral zones, the greater, in general, is the contribution of SD to S; in the polar caps within these zones, *SD* in general decreases as one proceeds to still higher latitudes, though there *SD* is usually still great compared with *Sq*.

The *Sq*, *L* and *D* fields change their form seasonally as the sun moves northward and/or southward across the equator. Also the *Sq* field and variation change somewhat irregularly from day to day—perhaps because of changing "weather" in the ionosphere, where Sq_e, the primary *Sq* current system flows. For brevity this account of the *D* field will refer mainly to the field as manifested when the sun is on the equator, that is, at the equinoxes.

The lunar daily variation and field *L*, at most places and times, are small compared with the *S* variation and field. Hence they have to be determined from considerable amounts of data, perhaps covering decades, and requiring much labor. In the discussion of *D* it is hardly necessary to take account of *L*. Hence we may take *D* to be given by the hourly values less the *Sq* inequality for each hour (determined as well as possible).

4.2. The morphology of the D field

Weak magnetic disturbance can best be studied statistically, and such methods are valuable also for magnetic storms, though these must also be studied individually. In the case of the *s*uddenly *c*ommencing (*Sc*) storms it is usually possible to determine the time of commencement to within about a minute, and time reckoned from this as origin is called storm time, and denoted by *T*. Even for storms that begin more gradually, it is often possible to determine the time of commencement to within about an hour, and thus *T* is known approximately also for such storms.

The *morphology* of a magnetic field is a term borrowed from the biologists, and signifies the space distribution of the field, as a function of time. The morphology of the *D* field at the earth's surface, during a magnetic storm, depends mainly on the space coordinates θ, λ, and on the storm time *T*. Here θ denotes the angle *PON* between the earth radii from the center *O* to the north pole *N*, and to *P* the point considered; λ denotes the longitude of *P* east of the Greenwich meridian. But more significant than λ is the longitude ϕ of *P* relative to the *midnight* meridian at the time considered. If the Greenwich time is *t*, so that *t*, reckoned in angle, at the rate 15° per hour, is the azimuth of the Greenwich meridian at that time, measured eastward from the midnight meridian, then

$$\phi = 15°t + \lambda, \text{ (t expressed in hour units).}$$

Thus ϕ is a coordinate of position relative to the midnight meridian, opposite to the noon or solar meridian; it is called the local time (in angular reckoning; in time reckoning, the local time is $\phi/15$).

Each magnetic storm is an individual, and differs from all others in some respects. But though the dependence of the *D* field on its main variables θ, ϕ and *T* is irregular, there are some broad features that are manifested by many, perhaps by most magnetic storms.

A geometrical division of the *D* field of a storm at any storm time *T* can be made as follows. For each element, e.g. for *H*, the *D* variation—in this case *D*(*H*)—can be expressed in the form

$$D(H) = Dst(H; T, \theta) + DS(H; T, \theta, \phi),$$

where the first part, Dst, represents the mean value of $D(H)$ round the parallel of latitude at polar angle θ; thus the second part, DS, has zero mean value round each such parallel. The D field is thus divided into two parts Dst and DS, such that the first part is symmetrical about the earth's axis, and the second part has zero mean value along each circle of latitude. Both parts can be determined in each element at each station for any storm at each storm time T. Both can be analyzed so as to determine their external and internal parts Dst_e, Dst_i and DS_e, DS_i. They are all functions of T and θ, and the DS fields are also functions of ϕ.

Figure 1 shows the average Dst changes in H, Z (or VF, the vertical force) and west declination, derived from 40 moderate magnetic storms. The Figure has three panels, left, middle and right; the left-hand panel is based on data from three low latitude stations, the middle one on four stations in about latitude 40°, the right-hand one on four stations with mean latitude 53°. The element most affected by Dst in these latitudes is H; $Dst(Z)$ is much smaller than $Dst(H)$, and opposite in sign; $Dst(d^0)$ is relatively negligible. The $Dst(H)$ curves show a first phase of increased H, lasting for a few hours, followed by a main phase in which H is reduced below normal, generally by much more than its increase during the first phase. The time interval in each panel of Fig. 1 is two days; the minimum of H is reached in about a day, and then there is a slow recovery towards normal, far from complete by the end of the second day of storm time.

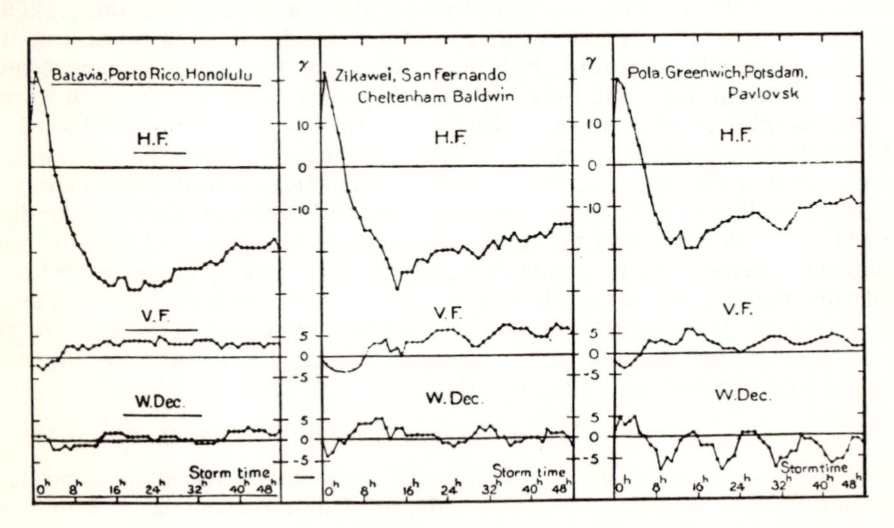

Fig. 4.1. Averaged storm-time magnetic disturbance-changes in different latitudes. (Chapman, 1918 p. 65; also Geomagnetism, p. 276.)

On the average, the more intense the storm, the more rapidly does it progress through its phases. This is shown by Fig. 2, which illustrates the average $Dst(H)$ changes during three groups of storms, weak, moderate and intense, during three days of storm time, in the mean for several low latitude stations.

The DS change along each latitude circle θ can be analyzed into Fourier components. The main one is the first, of form $c_1 \sin(\varphi + \epsilon_1)$; c_1 and ϵ_1 are functions of T and θ. Figure 15 shows also (by the curves marked *DS*) how c_1 varies with T, in the mean for the same low latitude stations. Its numerical value is shown as negative, merely for ease of comparison with *Dst*. It attains its extreme value earlier than *Dst*, and declines towards zero much more rapidly. The phase ϵ_1 also changes somewhat systematically with T.

Unlike *Dst*, the *DS* changes of vertical intensity Z and of W declination d^0 are comparable with $DS(H)$. Figure 3 shows $DS(H)$ and $DS(Z)$ for a series of stations or groups of stations, in latitudes up to 60° (the top curves), averaged over the first and second days of the storms on which Fig. 1 is based. The left-hand panels show $S(H)$ and $S(Z)$ for these stations, averaged over the *months* in which these storms occurred. The average of *DS* over a day represents *SD*; it is clear that the two uppermost $S(Z)$ curves contain a considerable *SD* part, whereas in the other cases $S(H)$ and $S(Z)$ have ranges comparable with those of *DS*, and the *S* curves differ in *type* from the *DS* curves; they represent *Sq* with good approximation.

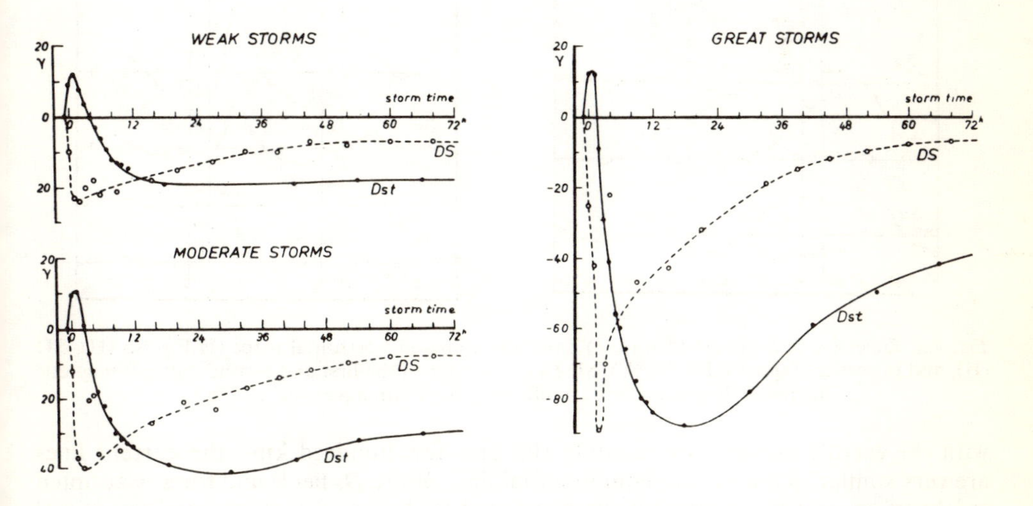

Fig. 4.2. A comparison of the rates of evolution of Dst(H) and the range ($2c_1$) of $DS_1(H)$ during the first three days of weak, moderate and great storms: mean dipole latitude 30°. Dst. curves are drawn with full lines, and those for DS with broken lines. (Sugiura and Chapman, 1960; p.45).

In auroral latitudes (60° upwards), beyond the latitudes to which Figs. 1 and 3 refer, the average *DS* variation during such storms is even greater than is shown in the top curves of Fig. 3. Its maximum ranges are found under or near the auroral zones, in approximately 67° magnetic dipole latitude. There it far exceeds *Dst*, although after decreasing in amplitude from the equator polewards, as shown in Fig. 1, $Dst(H)$ increases polewards towards the auroral zones. Within the auroral caps enclosed by the two auroral zones, northern and southern, *DS* is still considerable, and greater than *Dst*, though it is less than near the zones.

It is not easy to visualize the morphology of the *D* field from such sets of *Dst* and *DS* curves for different latitudes and elements, averaged over groups of magnetic storms. Help towards a clear conception is given by *conventional current diagrams*, showing electric current systems that *could* produce the observed surface distribution of the external part D_e of the *D* field. The word *conventional* is used to indicate that the diagrams do not necessarily indicate the current systems that *do* produce the surface D_e field. The conventional diagrams show a current system flowing at some constant height above the earth; if this height is small compared

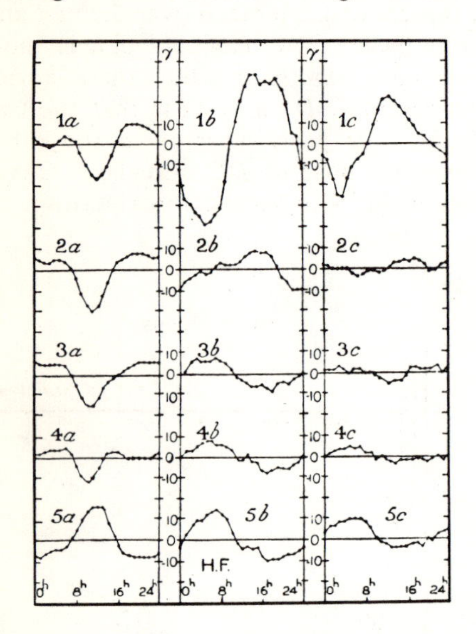

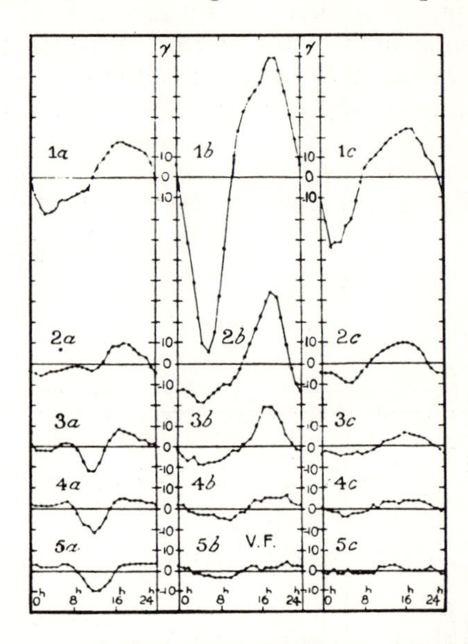

Fig. 4.3. Quiet (*a*) and disturbance (*b*, *c*) daily variations in horizontal force (H.F.), Sq (H), SD (H), and in vertical force (V.F.), Sq (V), SD (V); *b*, c refer to the first and second days of magnetic storms. (Chapman, 1918, p. 68, 69; also Geomagnetism, p. 278.)

with the earth's radius—for example 100 or a few hundred km—the current lines are very similar to the surface equipotential lines of the D_e field; and for any adopted height of current flow, they can be computed from the surface magnetic potential of the D_e field. Although the actual currents may only partly flow at a constant height, the diagrams easily indicate the surface distribution of the D_e field in all three components *X*, *Y*, *Z*.

Figure 4 shows such conventional current diagrams, partly constructed from the average *Dst* and *DS* curves for the moderate storms on which Figs. 1 and 3 are based. The diagrams in the left-hand panel indicate the currents over the sunlit hemisphere, as they would be seen, for example, from the sun. The right-hand panel shows the currents in the *N* hemisphere, as seen from above the *N* pole (magnetic axis latitude rather than geographic), with the midnight meridian pointing upwards. The current lines are drawn at intervals such that 10,000 amperes flow between adjacent lines. The middle diagrams show the currents that could produce the Dst_e part of the D_e field, and the lowest diagrams, those corresponding to the DS_e field. All the diagrams refer to the epoch of greatest development of the main phase of the storms. The top diagrams combine the current systems for Dst_e and DS_e and show the average D_e current system at that epoch. The strongest currents are those flowing along the auroral zone, mainly westward but also partly eastward. These currents are so concentrated that they are called (the auroral) *electrojets*. They complete their circuit mainly over the auroral cap enclosed by the zone, but also partly over the major belt of the earth between the two zones.

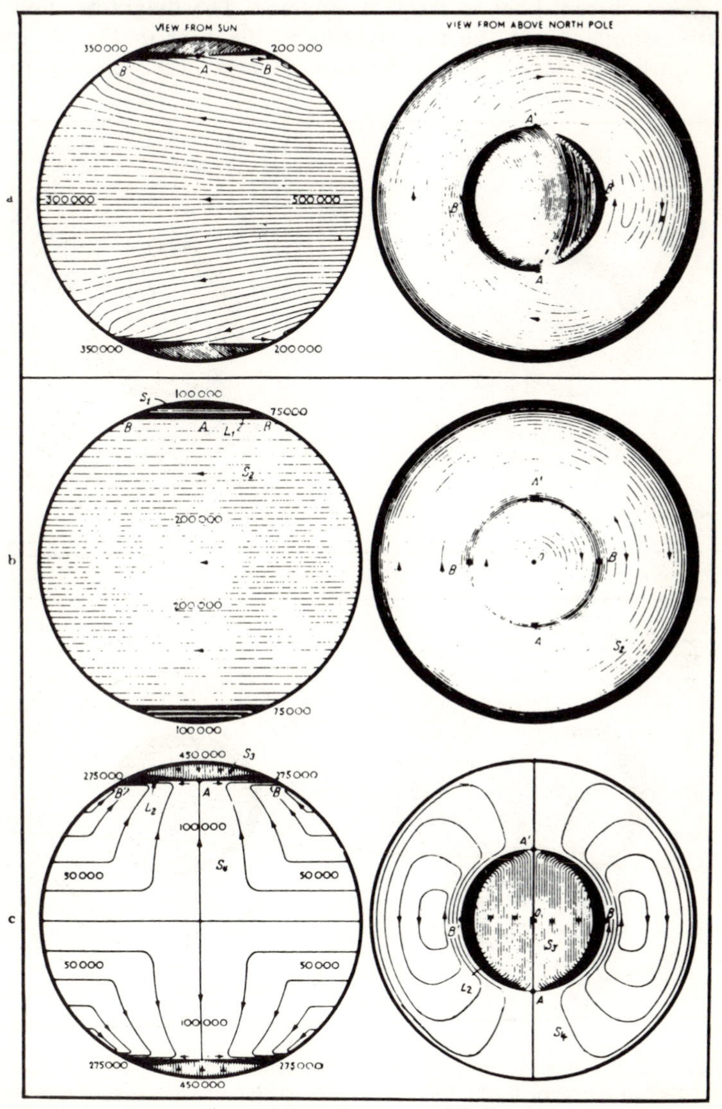

Fig. 4.4. Idealized current diagrams based on the average of forty moderate magnetic storms at maximum epoch in the main phase: (*a*) C(Dst), (*b*) C(DS), (*c*) C(D). The current flow between adjacent lines is 10,000 A. (Chapman, 1935, also Geomagnetism, pp. 302, 308, 310.)

The polar part of Fig. 4 is not based on the 40 magnetic storms already mentioned, but on miscellaneous data, partly from the (first) International Polar Year 1882/3. Figure 5 shows current diagrams later constructed by Vestine (1940) from data obtained during the (second) International Polar Year 1932/3. They refer to the *N* hemisphere as seen from above the magnetic axis pole. They are for four epochs during the moderate magnetic storm of May 1, 1933. The current flowing between adjacent current-lines is 100,000 amperes. The diagrams show considerable qualitative similarity to the top right-hand diagram in Fig. 4; quantitative equality cannot be expected, partly because storms differ greatly in intensity.

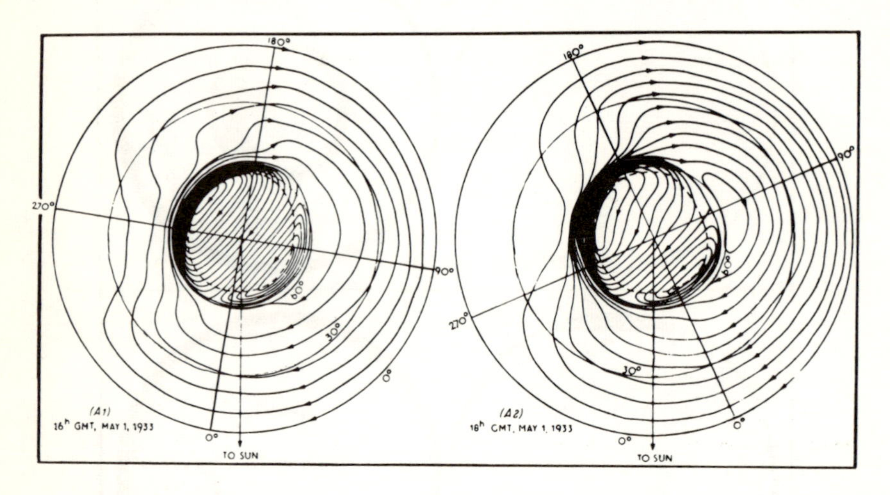

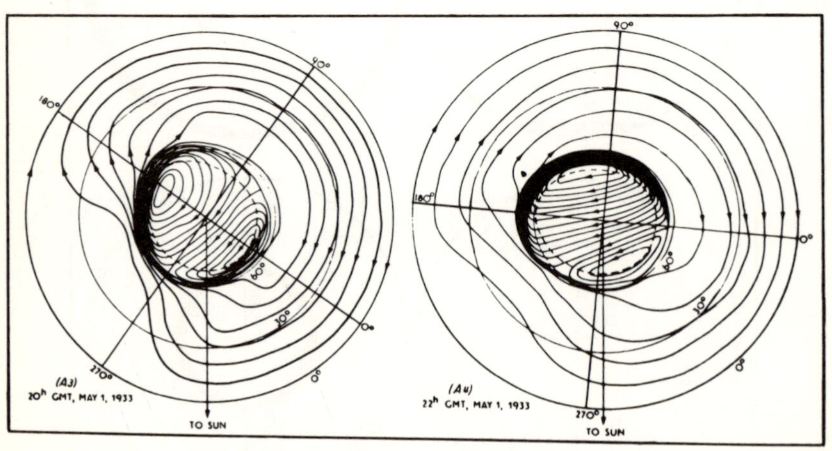

Fig. 4.5. Current diagrams for the main phase of a magnetic storm, 1st May, 1933, at four epochs. Current flow between adjacent lines: 100,000 A. The center of each diagram is the boreal axis pole *B*; the north geographical pole *N* lies 11° distant from *B* on the zero geomagnetic dipole meridian, as in Fig. 5. (Vestine, 1940, pp. 371, 372; also Chapman 1956a, p. 924, and also The Earth's Magnetism, p. 101.)

Some of the diagrams in Fig. 5 differ from the top right-hand diagram in Fig. 4 in that the eastward electrojet is weaker than in Fig. 4; and also in that the orientation of the auroral electrojets and auroral-cap currents is slightly or, in Fig. 5d, considerably different from that in Fig. 4.

These diagrams may be contrasted with those on page 15, which refer to the Sq_e field. There the pattern is very different; the auroral zones do not appear: the currents are weaker (of course the D_e conventional currents during weak magnetic disturbance may sink to similar or even less intensity). Another striking difference is that the Sq_e currents flow mainly over the sunlit hemisphere, and are weak over the night hemisphere. This feature does not appear in Figs. 4,5.

The Sq_e currents (also the L_e currents) flow in the lower ionosphere (E region), and are dynamo-induced, by daily varying air flow across the geomagnetic field. The ionization and electric conductivity of the E region are much greater by day than by night. Rocket observations have confirmed the inferences of this kind earlier made, as to the region of Sq_e and L_e current flow. Surface magnetic data *cannot* indicate the level of the currents, whose height was inferred from other evidence.

The location of the D_e currents is still not a unanimously agreed matter. However, the polar part of the conventional current system is generally thought to be ionospheric, flowing at about 100 km height: and likewise for part of the conventional currents between the two auroral zones, forming the balance of the return flow of the auroral electrojets. This ionospheric part of the D_e current system, and its field, are denoted by *DP* (*P* for polar). The *DP* field contributes most of *DS* and almost all the polar part of *Dst*.

But the part of the *Dst* field between the two zones, that is, over the major part of the earth's surface, probably originates far above the ionosphere (as this term is usually understood). The initial phase is ascribed to currents denoted by *DCF*

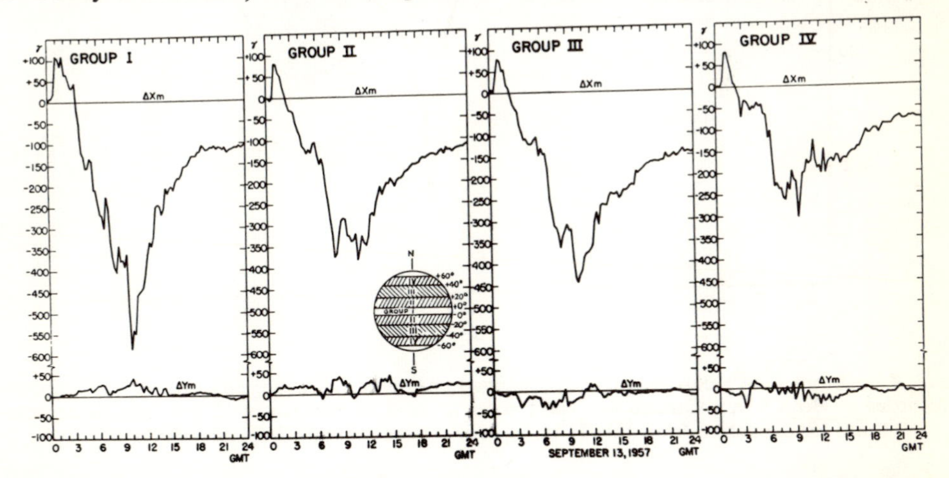

Fig. 4.6. The Dst(X_M) and Dst(Y_M) curves for September 13, 1957, for four zones bounded by circles of magnetic dipole latitude (indicated in the figure). The *Sc* occurred at 00h 47m GMT on September 13, 1957. Note the rapid growth and decay of the ring current. (Akasofu & Chapman, 1961, p. 1340.)

which depend on the solar Corpuscular Flux; they flow at the surface of the hollow in the solar plasma, as indicated by the Chapman-Ferraro theory. These currents flow as long as the corpuscular flux continues. Its duration is not easy to infer from the *Dst* curves. The main phase, of decreased *H*, is ascribed to a quite different "ring" current system, flowing in the radiation belts, probably within the *DCF* hollow. This current system and its field are denoted by *DR* (*R* for ring). It is likely that the plasma flow and its *DCF* field continue during much of the main phase of a storm, but that often the *DCF* field at the earth's surface is soon overpowered by the growth of the opposing *DR* field.

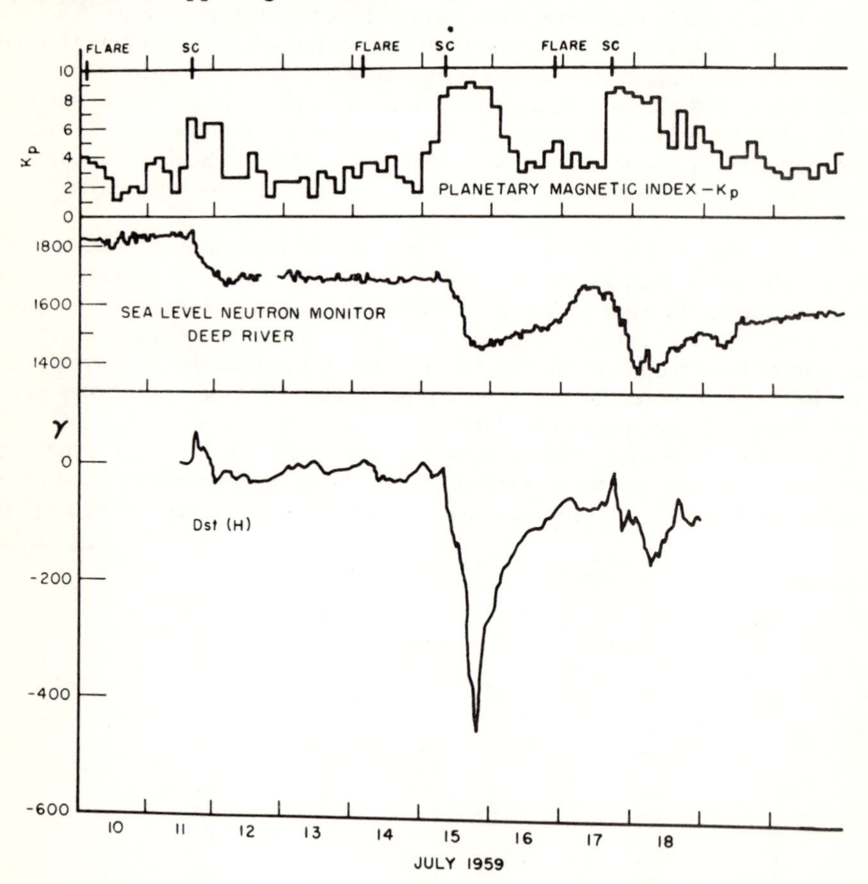

Fig. 4.7. (From above downwards) (*a*) Planetary magnetic index Kp. (*b*) Cosmic ray neutrons recorded at Deep River, Canada. (*c*) The Dst curve from the records of 12 low latitude observatories for the period from July 11 to 18, 1959. (Akasofu and Chapman, 1960a, p. 95.)

This account of the *D* field and its parts is thus far based on *statistical* studies of magnetic storms. It includes the *geometrical* subdivision of the *D* field, into an axially symmetrical part *Dst*, a function of θ and *T*, and the remaining *DS* field,

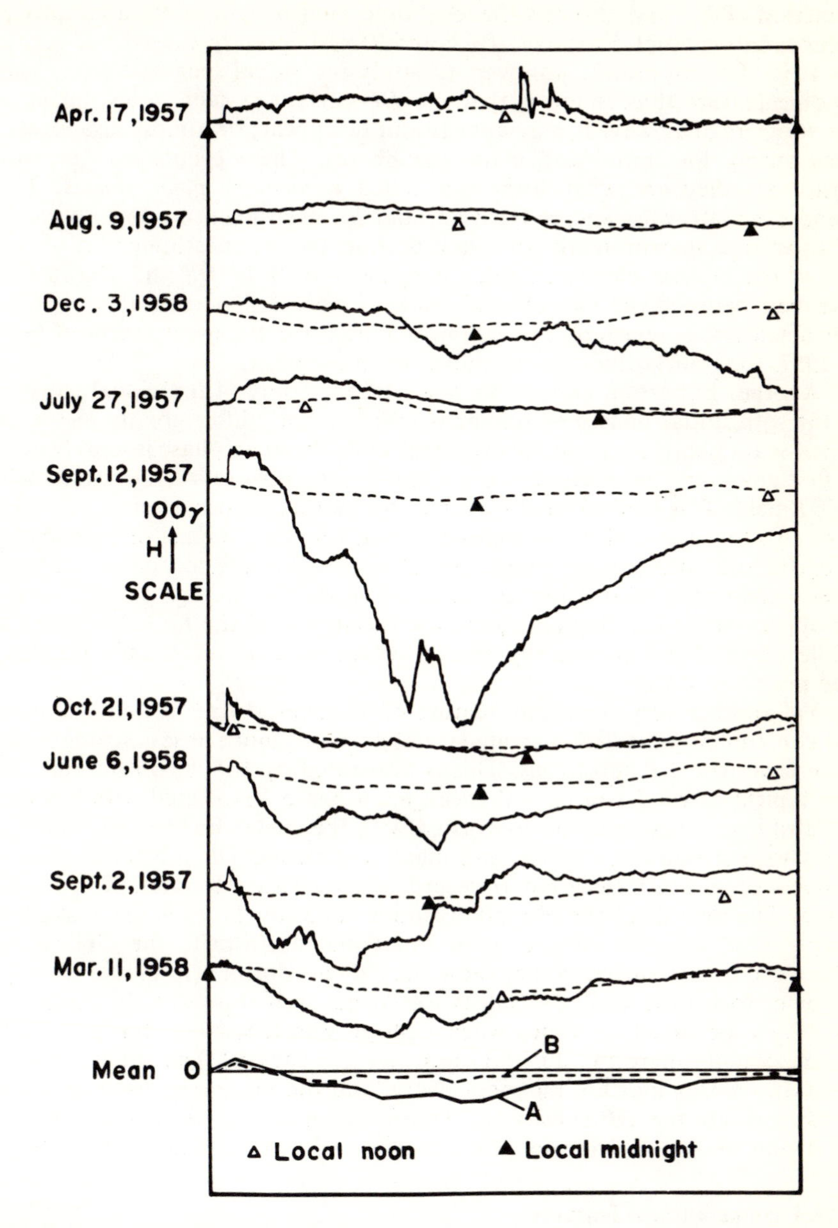

Fig. 4.8. Honolulu (Hawaii) and College (Alaska) records of horizontal force for contrasted magnetic storms; the hatched area in the Honolulu records shows the DR variation, and the blackened area in the College record shows the DP variations. (Akasofu and Chapman, 1962.)

a function of θ, T and also of φ the local time. It also includes the attempted *physical* division, into partial fields *DP*, *DCF* and *DR*, differently caused and located.

It is also important, however, to study individual magnetic storms. An outstanding feature thus found is that the *DP* current system varies during a storm in a very irregular way. It may appear and disappear, or almost disappear, several times during the initial and main *Dst* phases. These events may be called *DP* substorms; they are what Birkeland called *elementary polar storms*. They may extend over intervals ranging from an hour (or even less) to a few hours. As these *DP* substorms intermittently grow and decline, the lower latitude part of the return flow of the auroral electrojets adds irregularly to both *DS* and *Dst* in the major earth belt between the two auroral zones. Figure 6 illustrates this by *Dst* curves contructed (from relatively few station records) for the great storm of September 13, 1957, for four latitude zones, indicated in the inset.

Another important feature disclosed by the study of individual storms is that storms with initial phases of similar magnitude may differ greatly in the intensity of their main phase. As might be expected, when the main phase is weakly developed, the first phase may be prolonged for many hours, because it is not overwhelmed by the *DR* field. This is illustrated by three storms that all occurred in July 1959 (Fig. 7; see also Fig. 1.10); all three considerably affected the impact of cosmic rays upon the earth, and all three had similar initial phases; but the second storm was exceptionally great; the third storm also had a considerable though smaller main phase. The upper part of the diagram shows the variations of the *Kp* index (§1.9), and the middle curve shows cosmic ray effects, which were about equally notable in the three magnetic storms.

Yet another very significant feature of magnetic storms is that the main phase (and consequently the *DR* current system) develops more or less strongly, the more or the fewer the *DP* substorms. This is illustrated by Fig. 8; on the right it shows some typical cases of storms with weak main phase, associated with few and weak *DP* substorms; these may be contrasted with the curves on the left, which refer to storms with strong main phase, and many and strong *DP* substorms. For each of the five storms illustrated, the *H* record is shown for Honolulu and for College, Alaska. The records of the Honolulu station, in a low latitude (22°), show the *Dst* changes representative of the main interzonal earth-belt; the College records, taken close to the auroral zone, show the *DP* substorms. (It should be noted that for some storms the College records would not well show the *DP* substorms that might have occurred, at times when College was located in the part of the *DP* current system where the auroral electrojet is weakest.) The shaded parts of the Honolulu records indicate the main phase, and the blackened parts of the College records indicate the *DP* substorms. Those shown are all decreases, corresponding to a strong westward electrojet along the auroral zone, not far from College.

4.3. Some auroral features

The aurora is most frequently visible along the two auroral zones, which are oval, and centered approximately at the poles *B* and *A* (boreal and austral) of the magnetic dipole axis of the earth. The frequency of visibility decreases rapidly on the lower latitude sides of the zones, and less rapidly polewards within the auroral caps.

The auroral morphology (or changing distribution in space) is very complicated, and still imperfectly known. During the International Geophysical Year the auroral record depended largely on the work of about 120 all-sky cameras, of which about 90 were in northern latitudes, and about 30 in southern. They photographed the whole night sky visible from each station, when this was not obscured by clouds. As the auroral light comes mainly from 100 km height and above, each station could see all the auroras within a radius of about 700 miles or 1100 km: however, the details of the auroras that appeared at low elevations above the horizon were not very clear. In addition, visual observers in many countries gave valuable aid; they recorded auroras where these appeared less frequently, and where no all-sky cameras were used.

Auroral displays are even more variable and difficult to describe than is the *D* field; but some main features are here selected, as specially calling for theoretical explanation.

During a considerable magnetic storm that includes a few *DP* substorms, there will be a similar number of auroral displays, contemporaneous with the substorms. In general the displays start from or with a quiet phase, in which the aurora appears as one or more luminous arcs of rather simple form, lying approximately in the direction perpendicular to the magnetic dipole meridian. Suddenly there occurs the auroral "break up", when rays appear in the arcs, and the arcs become wavy and folded, and often move rapidly hither and thither over the sky. A great auroral fold on an arc during this active phase may cover a large part of the sky visible from one station; sometimes there are folds within folds. It is not yet certain whether nevertheless there is generally continuity of the surface of the arc. The arcs certainly often extend for thousands of km along the east-west direction. Their extent in height may range up to 100 km, or several hundred in the case of sunlit auroras that lie outside the shadowed atmosphere (though seen from places where the sun is visible). By contrast the thinness of the auroral arcs and bands is remarkable. In the quiet phase the arcs consist of long ribbons of light, with the great horizontal and vertical ranges just stated. But the thickness is only a few km, 3 to 5. In the active phase the luminous ribbons are even thinner, by a factor of order 10—namely 300 to 500 meters. These are among the outstanding facts of auroral morphology, which auroral theory must explain.

During very great storms the aurora comes within view of a much larger region of the earth than usual. For example, during the three greatest storms that occurred during the IGY, the aurora was seen from Mexico City in the tropics. One of these storms occurred on September 13, 1957; it is illustrated in Fig. 6. Another occurred on February 10/11, 1958; this storm seems to have been compounded of two, the second superposed on the first, with a new sudden commencement and initial phase; the latter started from a lower level (in *H*) than the first, because the recovery of *Dst*(*H*) was still incomplete. The second part of the storm was much the greater, and it was in this period that the aurora went to far lower latitudes than usual. It included four very intense *DP* substorms as it appears on the record of the Canadian observatory at Meanook. Between these substorms the auroral electrojet almost died away, and at these times the aurora could not be seen from Alaska, which usually is a region of frequent auroral visibility; the auroral arcs lay within a narrow band of sky of width only a few degrees of latitude. When the substorms were active, this band rapidly widened polewards and spread over an unusually great width in latitude, about 15°. Thus it was visible (in America) from Alaska to the southern USA (Akasofu and Chapman, 1962).

In the last phase of an auroral display the arcs and curtains disappear, and the sky may be covered with what look like dappled clouds faintly visible; but they appear and disappear rhythmically, showing that in fact they are auroral high level phenomena.

5. The Ring Current and its DR Field

5.1. Störmer's hypothesis concerning the DR field

In 1911/12 Störmer extended his auroral theory, which was based on charged solar particles moving in the earth's dipole field without influencing one another's motion. He knew that auroras appear in lower altitudes than usual during magnetic storms, when the earth's field is reduced. He sought to explain these two facts by the collective action of his particles—apparently it was the only case in which he seriously considered their combined action. Some of his electron orbits in the equatorial plane, namely those for $\gamma < -1$, sweep eastwards round the earth (beyond the radius C_{St}; see Fig. 3.5). Their motion corresponds to a westward current flowing partly near the earth. He suggested that this current might be the cause of what is here called the *DR* field (though his particles for which $\gamma > -0.5$ would correspond to an opposing, eastward current). He also showed that such a field would lower the latitude of entry of his particles into the atmosphere. For electrons of speed 10^8 cm/sec, his unit C_{St} is 5900 earth radii, or $\frac{1}{4}$ A.U.; thus the orbits would lie far beyond the moon. A westward current at this distance, sufficiently intense to reduce the earth's field by the amount observed during magnetic storms, would have a dipole moment enormously greater than that of the earth, and it would altogether change the field beyond about 10 earth radii—reversing the earth's field, which, however, was supposed to control the motion of these charges. The mutual repulsion of the particles, if present in numbers sufficient to explain the *DR* field, would disperse the particles. Thus his hypothesis was untenable.

It is discussed in his "Polar Aurora" (1955; see pp. 340–346), but with little numerical detail. His current did not encircle the earth, it was not a ring current. Hence it seems inappropriate to refer to the Störmerian ring current.

5.2. Schmidt's ring current

The true ring current was discussed on the basis of geomagnetic evidence by Schmidt (1916/7), the eminent long-time director of the Potsdam magnetic observatory. He concluded that the storm-time decrease of H must be caused by a current encircling the earth. He concluded also that it is present at all times, being enhanced during storms, and then dying away, but never completely. Thus he considered that it always makes some small (external) contribution to the M field, perhaps as much as 250γ. He seemed to locate the current above the atmosphere, as then understood, but he did not suggest any definite location or intensity. His conclusions might imply that some of the solar plasma whose impact on the earth is nowadays believed to cause magnetic storms becomes trapped in the earth's field, at least for a time, and has motions that constitute a ring current. But they would also be compatible with the alternative view, now considered, that the solar plasma is not itself trapped, but energizes atmospheric charged particles already present.

5.3. The Chapman-Ferraro ring current

Chapman and Ferraro (1932), after inferring the existence of the hollow carved by the geomagnetic field in the solar plasma, suggested that some of the gas gets trapped in the field. They doubtfully proposed a possible mechanism for the trapping, but did not stress their suggestion, or develop it quantitatively. It is still disputed whether solar matter is trapped, and if so, how. They assumed the existence of the ring current, however, from the magnetic evidence, and considered it likely to be of dimensions comparable with those of the hollow, hence of the order $10a$ (a = earth radius). They adopted a toroidal model for the current. The ions and electrons in this model circulate round the earth in a neutral stream, with the ionic motion, at least, westward, and enough difference between the ionic and electronic speeds to convey an adequate westward electric current. They considered the dynamic steady state of such a ring current, and showed that it was likely to have a duration of some days, at least, and that for certain types of perturbation it would be stable. They really only considered the motion of particles in the equatorial plane; particles not in this plane would necessarily have more complex paths.

Later Herlofson, according to Alfvén (1958), showed that for types of perturbation different from those considered by Chapman and Ferraro, this model ring current would be unstable.

5.4. Singer's model ring current

Singer (1957) proposed a different concept of the trapped component of the solar gas, based on the work of Störmer and Alfvén. He saw that the likely motion of the particles, and their distribution around the earth, would be of the kind not long afterward found in the "radiation" belts by satellite and cosmic rocket exploration (Van Allen and Frank 1959; Vernov *et al.* 1959). The particles oscillate rapidly to and fro between northern and southern mirror points in fairly high latitudes. At the same time they circle round the field direction, and also drift round the earth—the electrons eastward, the protons westward. Singer concluded that the presence of such particles, with these drift motions, would necessarily involve the existence of a westward ring current flowing round the earth. Later the ring current was studied in detail by Dessler and Parker (1959), Akasofu (1960), Akasofu with Chapman (1961) and also with Cain (1961), and by Apel and Singer (1961).

The Pioneer satellites I and III showed how variable is the main outer belt, at least as regards the particles of very high energy, which alone were recorded at that time. Later satellites have confirmed this variability, also for the more numerous particles of lower energy.

The ring current is here discussed specially in relation to the *steady state* of motion of a distribution of energetic charged particles in a dipole field. Such a discussion needs to be extended to consider the influence of the departure of the earth's field from that of a dipole: and also to consider varying states of the radiation belts, and of the motions of their particles.

5.5. The equation of motion of a charged particle

A particle of mass m and charge e (e.s.u.), moving in a magnetic field $\mathbf{B}$ and subject to an applied force $\mathbf{F}$ associated with a potential V, so that

$$\mathbf{F} = -\operatorname{grad} V, \tag{5.1}$$

has the following equation of motion (Larmor 1895, p. 724):

$$m\dot{\mathbf{v}} = (e/c)\mathbf{v} \times \mathbf{B} + \mathbf{F} = (e/c)\mathbf{v} \times \mathbf{B} - \operatorname{grad} V, \tag{5.2}$$

provided we ignore the loss of energy by the emission of electromagnetic radiation (see equation 15). If we multiply the terms of (2) scalarly by $\mathbf{v}$ we obtain the equation of energy:

$$(d/dt)(\tfrac{1}{2}mv^2 + V) = 0; \tag{5.3}$$

hence if $\mathbf{F} = 0$ the speed v of the particle is constant.

It is convenient to denote the components of $\mathbf{v}$ along and normal to $\mathbf{B}$ by u and w, so that

$$\dot{\mathbf{v}} = u\mathbf{b} + \mathbf{w}, \quad \text{where} \quad \mathbf{b} = \mathbf{B}/B, \quad \mathbf{b} \cdot \mathbf{w} = 0. \tag{5.4}$$

If θ is the angle between $\mathbf{v}$ and $\mathbf{B}$—it is called the *pitch* angle—

$$u = v\cos\theta, \quad w = v\sin\theta. \tag{5.5}$$

After dividing the terms of (2) by m, the equation may be written

$$\dot{\mathbf{v}} = \omega\mathbf{b} \times \mathbf{v} + \mathbf{F}/m, \tag{5.6}$$

where

$$\omega = -eB/mc. \tag{5.7}$$

Thus ω is positive if the charge is negative, and vice versa.

5.5a. Larmorian motion: B uniform and constant, F = 0

We first investigate the simplest case, in which $\mathbf{B}$ is uniform and constant, and $\mathbf{F} = 0$; the motion in this case is called Larmorian. The equation (6) reduces to

$$\dot{\mathbf{v}} = \omega\mathbf{b} \times \mathbf{v}, \tag{5.8}$$

where $\omega\mathbf{b}$ is a constant. This signifies that the velocity vector $\mathbf{v}$ simply rotates with constant angular velocity $\omega\mathbf{b}$; thus the component velocity $u\mathbf{b}$ along $\mathbf{B}$ is constant, and the transverse velocity component $\mathbf{w}$ rotates with this angular velocity. Its components along two fixed directions $\mathbf{i}$, $\mathbf{j}$ normal to $\mathbf{b}$ and to each other, and such that $\mathbf{b}$, $\mathbf{i}$, $\mathbf{j}$ is a right-handed triad, will be

$$w\cos\alpha(= v\sin\theta\cos\alpha), \quad w\sin\alpha(= v\sin\theta\sin\alpha), \tag{5.9}$$

where

$$d\alpha/dt = \omega. \tag{5.10}$$

The rotation, viewed looking along the direction of **B**, is positive (clockwise) if ω is positive (that is, for a negative charge), and anticlockwise for positive e. The values of v, u, w and θ are arbitrary but constant throughout the motion. The path, a helix of ("Larmor") radius R, has the equation, obtained by integration of (8):

$$\mathbf{r} = \mathbf{r}_0 + u\mathbf{b}t + R(\mathbf{i}\sin\alpha - \mathbf{j}\cos\alpha), \tag{5.11}$$

where $\mathbf{r}_0$ denotes the (arbitrary) initial position vector of the particle, and

$$R = w/\omega = mcw/eB. \tag{5.12}$$

The helical frequency $\omega/2\pi$, here called the *Larmor* or *gyro* or *cyclotron* frequency, is

$$eB/2\pi mc \tag{5.13}$$

(The precessional frequency of an orbital electron, ω/π, is also named after Larmor). Larmor (1897, p. 512) showed that a charged particle moving with acceleration $\dot{\mathbf{v}}$ would radiate electromagnetic energy at the rate P given by

$$P = 2e^2\dot{v}^2/3c^3. \tag{5.14}$$

In the present case $\dot{\mathbf{v}} = \dot{\mathbf{w}}, |\dot{\mathbf{v}}| = |\dot{\mathbf{w}}| = |\omega|w = |e|wB/mc$, so that

$$P = 2e^4w^2B^2/3m^2c^5 = f(\tfrac{1}{2}mw^2), \tag{5.15}$$

where

$$f = 4e^4B^2/3m^3c^5. \tag{5.16}$$

For an electron this indicates a time of decay of the transverse kinetic energy of the particle in $2.55 \times 10^8/B^2$ seconds; for a field such as that of the earth this time is very long; hence it is justifiable to ignore the corresponding retarding term in the equation of motion (2). Alfvén (1950), however, pointed out that though this is true for the motion of a single particle, the radiative reaction may be important when many particles are present, if their effects on each other are cooperative.

The electromagnetic radiation at the rate P is determined by the combined electric and magnetic fields associated with the moving particle. These are most simply calculated with reference to axes sharing the mean motion $u\mathbf{b}$ of the particle Relative to this frame there is an electrostatic field corresponding to a charge e at the mean position of the particle, namely at the center of the Larmor circle. There are also electric and magnetic fields varying with the Larmor frequency, and due to the rotating electric dipole composed of a balancing charge $-e$ at the center of the Larmor circle, and the actual revolving charge e. At any point the mean of the varying electric field is zero; the mean magnetic field is that of a uniform electric current J flowing round the Larmor circle, where J corresponds to the circulation, with speed w, of a uniform line density of charge equal to that of the charge e spread uniformly round the circle, namely $e/2\pi R$. Thus $J = |e|w/2\pi cR$ in e.m.u. The magnetic moment **M** of this current flow has the direction $-\mathbf{b}$, and its magnitude is $\pi R^2 J$; thus

$$\mathbf{M} = -(mw^2/2B)\mathbf{b}. \tag{5.17}$$

The magnetic flux through the Larmor circle is $\pi R^2 \mathbf{B}$ or

$$-(2\pi mc^2/e^2)\mathbf{M}. \tag{5.18}$$

5.5b. Modified Larmorian motion: B uniform and constant, F constant

When **B** is uniform and constant, and there is a constant force **F**, the solution of the equation of motion (6) is the sum of a part which is the solution of (8)—this is the complementary function, in the terminology of differential equations—and of a part that is any solution of (6)—the particular integral, the simpler the better. Such a particular integral is

$$F'\mathbf{b}t/m + (c/eB)\mathbf{F} \times \mathbf{b}, \quad \text{where} \quad F' = \mathbf{F}\cdot\mathbf{b} = F\cos\theta', \tag{5.19}$$

if θ' is the angle between **F** and **B**. This corresponds to uniformly accelerated motion along the field, due to the F' component of **F**, together with a uniform "drift" velocity perpendicular both to **F** and to **B**, with speed $(c/|e|B)F\sin\theta'$. The total motion consists of a "mean" motion

$$(u_0 + F't/m)\mathbf{b} + (c/eB)\mathbf{F} \times \mathbf{b}, \tag{5.20}$$

together with the Larmorian circular motion, as if **F** were zero. The resultant motion in the plane normal to **B** is trochoidal.

5.5c. Motion in a constant non-uniform magnetic field

When the magnetic field is not uniform, the exact solution of (2) is difficult, even when **F** is zero. In general the motion is determinable only by numerical computation. Störmer throughout several decades organized such computations for the case of a dipole field (with **F** zero). The numerical labor increases with the number of Larmor loops described; this depends on the particle parameters, which in the dipole case include the Störmer length unit C_{St} given by (3.65), in which M denotes the *field* dipole magnetic moment; they also include the initial position and velocity.

Alfvén (1940) discussed the general problem from the standpoint of perturbation theory, valid when the Larmor radius (12) at each position of the particle is small compared with the scale length L of the field, and—if the field is varying—when the Larmor time of revolution $2\pi/\omega$ is small compared with the scale time T of the variation; L is of order $1/\text{grad} \ln B$, and T is of order $dt/d \ln B$, where ln signifies the Napierian logarithm. Alfvén considered separately, as perturbations, the influences of (a) a weak force **F**, (b) a slight non-uniformity of the field, and (c) a slow variation of the field. He was thus led to the conception of the motion as that of approximately Larmorian motion, with slowly varying parameters ω and R, relative to a smooth motion of a point that he called the guiding center. Many discussions of the motion, like the original discussion by Alfvén, use a separate fairly simple argument for each part of the motion of the guiding center. As Fried (1960) has remarked, this procedure may leave the reader in doubt as to whether some term may have been omitted, or perhaps counted twice. The objective is a solution expressed as an expansion in powers of some parameter of the motion, such as $1/\omega$, valid when this is small enough. Poincaré remarked, relative to such perturbation calculations, that the division of the series is not unique; different choices can be made. This arbitrary feature of the solution can be removed by formulating some convenient principle(s) of choice.

Such a method of obtaining an asymptotic series solution of (2) was given by Bogolyubov and Zubarev (1955); their paper was translated into English by Fried (1960). Their solution is here given without proof (the analysis is somewhat long), for a constant field **B**, but only up to terms of order $1/\omega$.

The solution is expressed relative to a right-handed triad of unit vectors at each position P occupied by the particle, namely

$$\mathbf{b}, \mathbf{c}, \mathbf{d}, \quad \text{where} \quad \mathbf{d} = \mathbf{b} \times \mathbf{c}; \tag{5.21}$$

b is given by (4), and **c** is the unit vector in the direction of CP, where C denotes the center of curvature of the **B** field line at P. Thus **b** is along the tangent to the field line at P, **c** is along the principal normal, and **d** is along the binormal; their directions may change from point to point along the line. By definition of the center C and the radius ρ of curvature at P,

$$(\mathbf{b}\cdot\nabla)\mathbf{b} = -\,\mathbf{c}/\rho, \tag{5.22}$$

because $(\mathbf{b}.\nabla)\,\mathbf{b}$ is the rate of variation of **b** along the field line, at P.

The solution is expressed in terms of the position $\bar{\mathbf{r}}$ and velocity $\bar{u}\mathbf{b} + \bar{\mathbf{w}}$ of the guiding center, and of the oscillatory motion relative to this center. The displacement $\mathbf{r} - \bar{\mathbf{r}}$ is of order $\bar{\mathbf{w}}/\omega$ or R, and the differences $u - \bar{u}$, $\mathbf{w} - \bar{\mathbf{w}}$ are fractions of $\bar{u}$ and $\bar{w}$, of order $1/\omega$; in fact

$$\mathbf{r} = \bar{\mathbf{r}} + \bar{R}(\mathbf{c}\sin\bar{\alpha} - \mathbf{d}\cos\bar{\alpha}), \tag{5.23}$$

where $\bar{\alpha}$ differs from the α of equation (9) by two terms, both small (in the circumstances considered) and periodic respectively in $\bar{\alpha}$ and $2\bar{\alpha}$. Similarly $u - \bar{u}$ and $w - \bar{w}$ each consist of two such periodic terms.

The equation of motion, to the zeroth approximation (neglecting terms with additional factors that are powers of $1/\omega$, and therefore ignoring the difference between u and $\bar{u}$, w and $\bar{w}$), leads to the component equations:

$$\dot{u} = F' + \tfrac{1}{2}w^2 \operatorname{div}\mathbf{b}, \quad \dot{w} = -\,\tfrac{1}{2}uw \operatorname{div}\mathbf{b}. \tag{5.24}$$

Here

$$\operatorname{div}\mathbf{b} = \operatorname{div}(\mathbf{B}/B) = (1/B)\operatorname{div}\mathbf{B} + \mathbf{B}\cdot\operatorname{grad}(1/B) = \mathbf{B}\cdot\operatorname{grad}(1/B) \tag{5.25}$$

because $\operatorname{div}\mathbf{B} = 0$. Hence

$$\begin{aligned} \dot{w} &= -\,\tfrac{1}{2}uw\mathbf{B}\cdot\operatorname{grad}(1/B) = -\,\tfrac{1}{2}Bw\mathbf{u}\cdot\operatorname{grad}(1/B) \\ &= -\,\tfrac{1}{2}Bw\mathbf{v}\cdot\operatorname{grad}(1/B) = -\,\tfrac{1}{2}Bw(d/dt)(1/B), \end{aligned} \tag{5.26}$$

because **u** and **v** differ only by a fraction of order $1/\omega$. The integral of (26) is

$$w^2/B = \text{constant}. \tag{5.27}$$

This integral applies to the value of w^2/B along the path of the particle. Thus w^2/B is an adiabatic invariant of the motion (correct if we ignore terms smaller by factors of order $1/\omega$). As shown in §5.5a, this means that the diamagnetic dipole moment of the Larmorian motion, and the field flux through the Larmorian circle, are constant throughout the motion (see 17, 18). It implies that the transverse speed w of the particle increases as the particle moves into a stronger part of the field, e.g., when a particle in the geomagnetic field moves earthward from the equatorial plane towards the northern or southern hemisphere. If $\mathbf{F} = 0$, so that the total speed v is constant, any gain to w must be at the expense of the component speed u along the field, and must involve an increase of the pitch angle θ. The maximum value of θ is 90°, corresponding to $w = v$, $u = 0$; a point where these values of θ, w, u are reached is called a *mirror* point, because there the forward motion into the stronger field is stopped and reversed.

The invariant equation (27) is valid also when the relativity variation of mass is taken into account, and when the field B and the force $\mathbf{F}$ (for example, due to an electric field) are changing, provided that the Larmor time of circular motion is small compared with the scale time of variation of the fields (Berkowitz and Gardner, 1957; Kruskal 1960; Northrop 1961). The $\mathbf{b}$ component of any electric field present must be suitably small.

5.6. The drift motion

The next approximation to the solution of (2) includes terms containing $1/\omega$ as a factor; they represent drift motion additional to that due to $\mathbf{F}$. Northrop (1961; his equation 17) gives the full expression to this degree of approximation, including terms due to an electric field $\mathbf{E}$, which contributes all or part of $\mathbf{F}$, and taking both $\mathbf{B}$ and $\mathbf{E}$ to vary with time. Here the result will be given only for the case of $\mathbf{B}$ constant, and $\mathbf{F} = 0$; it can be expressed in several different ways, e.g., denoting the position vector of the guiding center by $\bar{\mathbf{r}}$,

$$\dot{\bar{\mathbf{r}}} = u\mathbf{b} + (mcw^2/2eB^2)\mathbf{b} \times \nabla B + (mcu^2/eB)\mathbf{b} \times (\mathbf{b}\cdot\nabla)\mathbf{b}. \qquad (5.28)$$

The second term on the right comes from the non-uniformity of the $\mathbf{B}$ field; the third is the "acceleration" drift, due to the centrifugal force $mu^2\mathbf{c}/\rho$ associated with the motion of the guiding center along the curved line of force (cf. 22).

5.7. The drift current; B non-uniform and constant; F = 0

The drift motions of an assemblage of charged particles convey an electric current, the drift current. This current is here considered for a *neutral* ionized gas, in which there is no electric field or other applied force. The magnetic field $\mathbf{B}$ is non-uniform but constant. The density is taken to be low enough for the particles not to affect one anothers' motions, except through their collective modification of the magnetic field. Thus (28) gives the motion to be considered; the first term, $u\mathbf{b}$, is not regarded as a drift, and it is supposed that the distribution of values of u is symmetrical with respect to sign, so that $u\mathbf{b}$ gives no resultant electric current along the lines of force. Thus the drift current $\mathbf{i}_{dr}$ is the sum, over all the charged particles, of $e\,\dot{\bar{\mathbf{r}}}$, namely

$$\mathbf{i}_{dr} = \sum\{(mcw^2/2B)\mathbf{b}\times\nabla B - (mcu^2/B^2\rho)\mathbf{d}\} \qquad (5.29)$$

after substitution from (22, 21).

This can be expressed in terms of the ionized gas pressures p_b, p_n along and normal to **B**, given by

$$p_b = \sum \int mu^2 n(\mathbf{v}) d\mathbf{v}, \quad p_n = \sum \int \tfrac{1}{2} mw^2 n(\mathbf{v}) d\mathbf{v}. \tag{5.30}$$

The sum symbol refers to the different kinds of charged particle, e.g., electrons and protons; the integrals are to be taken throughout the velocity space, of which $d\mathbf{v}$ denotes one element, and $n(\mathbf{v})$ denotes the velocity distribution function. The factor $\frac{1}{2}$ in the expression for p_n is required because w^2 is the sum of the squares of its two components (9). Thus (29) may be replaced by

$$\mathbf{i}_{dr}/c = (p_n/B^2)\mathbf{b} \times \nabla B - (p_b/B\rho)\mathbf{d}. \tag{5.31}$$

5.8. Diamagnetism and the equivalent electric current

The circular Larmorian motion of the particles around their guiding centers gives each of them the magnetic moment **M**. As $\mathbf{M} = -(mw^2/2B)\mathbf{b}$, the gas is diamagnetic, with intensity **I** of magnetization, given by

$$\mathbf{I} = -(p_n/B)\mathbf{b} = -(p_n/B^2)\mathbf{B}. \tag{5.32}$$

A distribution of magnetization of intensity **I** is equivalent to a distribution of electric current of intensity c curl **I** (Stratton 1941, p. 237). If the distribution of magnetization is discontinuous at the boundary, the volume current distribution is supplemented by a surface current distribution, of surface intensity $\mathbf{I} \times \mathbf{n}$, where **n** denotes the unit normal vector outward from the magnetized volume; together the volume and surface currents form a non-divergent system (Stratton 1941; p. 242). Here it will be supposed that the gas density has no sharp boundary, so that the surface current will not be present.

Let $\mathbf{i}_{di}$ denote the electric current intensity equivalent to the diamagnetic distribution (32). Then

$$\begin{aligned} \mathbf{i}_{di}/c &= \operatorname{curl} \mathbf{I} = -\operatorname{curl}(p_n \mathbf{B}/B^2) \\ &= -(p_n/B^2)\operatorname{curl}\mathbf{B} + (\mathbf{b} \times \nabla p_n)/B - (2p_n/B^2)\mathbf{b} \times \nabla B \end{aligned} \tag{5.33}$$

5.9. The total plasma current intensity i

The total plasma current intensity **i**, the sum of the drift current $\mathbf{i}_{dr}$ and the diamagnetism equivalent current $\mathbf{i}_{di}$, is given by combining (31, 33), as follows:

$$\mathbf{i}/c = (\mathbf{b} \times \nabla p_n)/B - (p_b/B\rho)\mathbf{d} - (p_n/B^2)\mathbf{b} \times \nabla B - (p_n/B^2)\operatorname{curl}\mathbf{B} \tag{5.34}$$

The last term but one in (34) may be transformed as follows:

$$\mathbf{B} \times \operatorname{curl}\mathbf{C} = \mathbf{B} \times (\nabla \times \mathbf{C}) = \nabla_c(\mathbf{B}\cdot\mathbf{C}) - (\mathbf{B}\cdot\nabla)\mathbf{C}. \tag{5.35}$$

Here the subscript C signifies that the operator ∇ there acts only on **C**. Interchanging **B** and **C** in (35), adding, and making $\mathbf{C} = \mathbf{B}$, we obtain:

$$\mathbf{B} \times \operatorname{curl}\mathbf{B} = B\nabla B - (\mathbf{B}\cdot\nabla)\mathbf{B} = B\nabla B - B^2(\mathbf{b}\cdot\nabla)\mathbf{b} - \mathbf{B}(\mathbf{b}\cdot\nabla)B \tag{5.36}$$

Vector multiplication by $\mathbf{b}/B$ then gives the equation

$$\mathbf{b} \times \nabla B = (\mathbf{b}\cdot \text{curl}\, \mathbf{B})\mathbf{b} - \text{curl}\, \mathbf{B} - (B/\rho)\mathbf{d}. \tag{5.37}$$

Hence (34) may be transformed to:

$$\mathbf{i}/c = (\mathbf{b} \times \nabla p_n)/B + (p_n - p_b)\mathbf{d}/B\rho - (p_n/B^2)(\mathbf{b}\cdot \text{curl}\, \mathbf{B})\mathbf{b}. \tag{5.38}$$

If there is an electrical field $\mathbf{E}$ everywhere transverse to $\mathbf{B}$, the resulting drift motion $(c/B)\mathbf{E} \times \mathbf{b}$ given by (19) is the same for positive and negative charges. Hence it makes no contribution to the electric current.

In (38) curl $\mathbf{B}$ may be replaced by $4\pi\mathbf{i}/c$; scalar multiplication of all terms of (38) by $\mathbf{b}$ leads to

$$(1 + 4\pi p_n/B^2)(\mathbf{i}\cdot \mathbf{b}) = 0. \tag{5.39}$$

The first factor in (39) may be expressed as $1 + p_n/2p_m$, where p_m denotes the magnetic pressure $B^2/8\pi$. As this factor is not zero, $\mathbf{i} \cdot \mathbf{b} = 0$ or $\mathbf{b} \cdot \text{curl}\, \mathbf{B} = 0$. Consequently (38) reduces to

$$\mathbf{i}/c = (p_n - p_b)\mathbf{d}/B\rho + (\mathbf{b} \times \nabla p_n)/B \tag{5.40}$$

5.10 The pitch angle and density distributions

The velocity distribution function $n(\mathbf{v})$ at any point P in the field is a function of the speed v and the pitch angle θ, but not of α, the azimuthal coordinate in (9); this is because of the Larmor rotation of the transverse velocities $\mathbf{w}$ of the particles. The function $n(\mathbf{v})$ may also depend on the position of P, because the magnetic control of the motions of the particles determines the density and the velocity distribution. Parker (1957) showed that the space variation of $n(v)$ is specially simple if $n(v)$ involves θ only by a factor $\sin^{\beta+1}\theta$, because if this is the case at one point of a line of force, it is so all along the line. Conceivably β might vary from one line of force to another, or depend on the speed v, or on the kind of particle. Here, for simplicity, β will be supposed constant throughout the field. Let $n(v)$ denote the *speed* distribution function, so that $n(v)dv$ is the number of particles per unit volume with speeds between v and $v + dv$. Of these, the number

$$A(\beta) \sin^{\beta+1}\theta n(v)dvd\theta \tag{5.41}$$

have pitch angles between θ and $\theta + d\theta$. The factor $A(\beta)$, given by

$$A(\beta) = \Gamma(\beta + 2)/2^{\beta+1}\{\Gamma(\tfrac{1}{2}\beta + 1)\}^2, \tag{5.42}$$

is chosen so that

$$\int_0^\infty A(\beta)\sin^{\beta+1}\theta d\theta = 1, \qquad \int n(\mathbf{v})d\mathbf{v} = \int^\infty n(v)dv = n,$$

where n denotes the total number density (for any one kind of particle present). With such a pitch angle distribution, $n(v)$ varies along the line as $B^{-\rho/2}$, so that if it is $n'(v)$ at a point P' where the magnetic intensity is B', $n(v)$ at P is given by

$$n(v) = (B'/B)^{\beta/2}n'(v). \tag{5.43}$$

When $\beta = 0$ the velocity direction distribution is uniform or "spherical", because $\sin \theta d\theta$ is proportional to the area of a zone of a sphere between the polar angles θ and $\theta + d\theta$. In this case $n(v)$ is the same all along any line of force.

When $\beta \neq 0$ the velocity direction distribution is non-uniform; the surface density of velocity points for particles of speed v, on the velocity sphere of radius v, is proportional to $\sin^{\beta}\theta$; for $\beta > 0$ the larger pitch angles or ratios w/u are in excess, and for $\beta > 0$, the smaller pitch angles. Figure 1 illustrates this surface

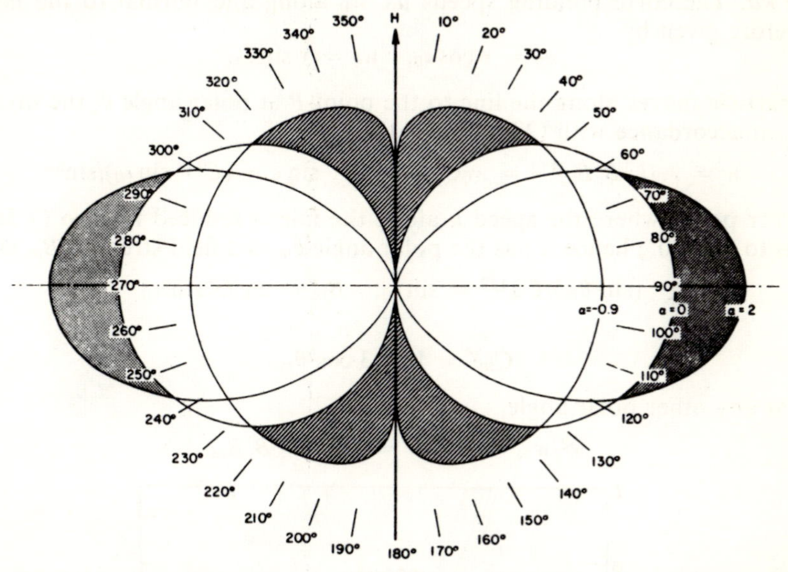

Fig. 5.1. The pitch-angle distribution $F \propto \sin^{\beta+1} \theta$ for $\beta = 2$, 0, and -0.9. The direction $\theta = 0$ is parallel to H. The curves are normalized to give the same mean density in each case. The deviation from the isotropic distribution ($\beta = 0$) is shown by a dotted area for $\beta = 2$ and by a hatched area for $\beta = -0.9$. For $\beta > 0$ the larger pitch angles are in excess; for $\beta < 0$, the smaller pitch angles. (Akasofu and Chapman, 1961, p. 1326.)

density for three values of β, namely -0.9, 0 and 2; the curves are normalized to give the same mean density in each case. According as $\beta > 0$ or $\beta < 0$ the number density increases or decreases with increasing θ.

5.11 The motion of the charges in a constant dipole field, F = 0.

The preceding analysis is here applied to a constant dipole field, for the case $\mathbf{F} = 0$; thus any particle moves with constant speed v. The dipole field is as described in § 1.2 and illustrated in Fig. 1.3; it is convenient to take the radius a of § 1.2 to be the earth's radius (6.37×10^8 cm), and the moment M and the equatorial intensity H_0 at the surface of the sphere ($r = a$) to be given by (1.20). To avoid confusion with the polar angle θ of § 1.2, the pitch angle of the particle motion (equations (5.9) and § 10) is here denoted by ϵ.

The **B** field lines have the equation (1.10); the factor k is here taken as the parameter of the line. The line crosses the equatorial plane at $r = ka$, its greatest distance from O. Its radius of curvature ρ at the point P on this line at polar angle θ is given by (1.13); at that point

$$B = H_0 C/k^3 \sin^6\theta = B_0 C/k^3 \sin^6\theta \tag{5.44}$$

(cf. 1.8, 1.10, where F and H_0 denote the quantities here written B and B_0); C is given by (1.12), and B_0 is the equatorial value of B at $r = a$.

Let ϵ_0 be the pitch angle of a particle where it crosses the equatorial plane at distance ka. The corresponding speeds u_0, w_0 along and normal to the field line are therefore given by

$$u_0 = v \cos \epsilon_0, \quad w_0 = v \sin \epsilon_0. \tag{5.45}$$

As the particle moves along the line to the point P at polar angle θ, the invariance of w^2/B, in accordance with (27), implies that there

$$w = w_0(k^3 B/B_0)^{1/2} = w_0 C^{1/2}/\sin^3\theta, \quad \sin \epsilon = (C^{1/2} \sin \epsilon_0)/\sin^3\theta \tag{5.46}$$

The mirror point, where the speed u along the line is reduced to zero (§ 5c), corresponds to $\epsilon = 90°$; hence it has the polar angle θ_m and field strength B_m given by

$$(\sin^3\theta_m)/C_m{}^{1/2} = \sin \epsilon_0, \quad B_m = B_0/k^3 \sin^2\epsilon_0 \tag{5.47}$$

where

$$C_m{}^2 = 1 + 3 \cos^2\theta_m. \tag{5.48}$$

Hence, at any other polar angle,

$$w^2/w_m{}^2 = w^2/v^2 = \sin^2\epsilon = B/B_m. \tag{5.49}$$

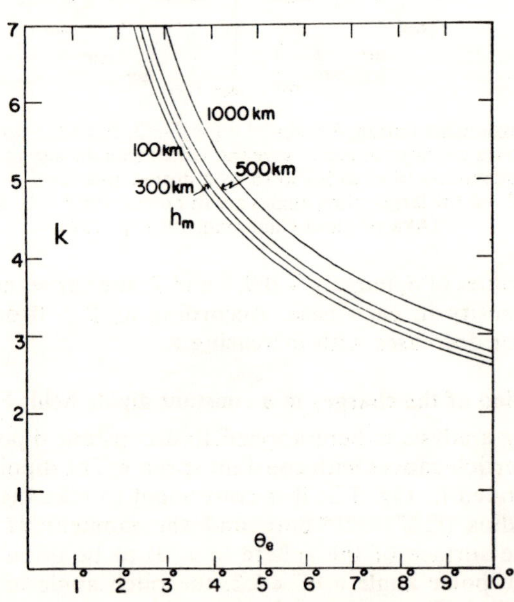

Fig. 5.2. The graphs show for what combinations of k and ϵ_0 the mirror points have heights of 100, 300, 500, and 1000 km above the ground; ka is the distance at which a particle crosses the equatorial plane, with pitch angle ϵ_0. (Akasofu and Chapman, 1961, p. 1325.)

The mirror point is at the corresponding radius $r_m = ka \sin^2\theta_m$, at a height h_m above the ground ($r = a$). Figure 2 illustrates by graphs how h_m depends on the parameter k of the line of force and on the equatorial pitch angle ϵ_0. For example, Fig. 2 shows that in order to approach the earth within 100 km, a particle moving on the surface of revolution formed by the lines of force with parameter $k = 6$—the surface that reaches the earth at polar angle 24°.1 (see Table 1.1) in the auroral zone—must have an equatorial pitch angle less than 3°: that is, it must cross the equatorial plane almost perpendicularly, almost along the line of force. This requirement ignores the resistance of the atmosphere; a particle that can penetrate the atmosphere to normal auroral levels (about 100 km) would in a vacuum have a mirror point corresponding to a smaller (actually negative) value of h_m—implying a still smaller value of ϵ_0.

Note that the relation (47) between θ_m and ϵ_0 is independent of the speed of the particle.

Auroras that occur in such an unusually low latitude as 45° (where also $\theta = 45°$) correspond to entry of the particles along the line of force with parameter $k = 2$ (Table 1.1). For such particles to approach the earth's surface (*in vacuo*) to within 100 km, the limiting maximum pitch angle is considerably greater, namely arc sin $(160)^{-\frac{1}{4}}$ or 16°.3.

At the mirror point the particle motion in latitude is reversed, and it returns to cross the equator and reach a corresponding mirror point on the other side. The time T_0 required for a complete oscillation in latitude, from the equator to a mirror point, thence to the further mirror point, and back to the equator again, is clearly given by

$$T_0 = 4s/v, \qquad (5.50)$$

where s denotes the distance along the *spiral path*, from the equator to either mirror

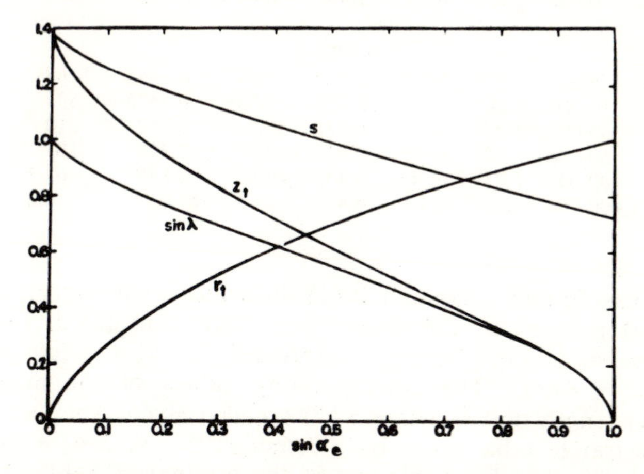

Fig. 5.3. Various quantities are plotted as functions of the equatorial pitch angle ϵ_0 of a trapped particle; s is the dimensionless arc length of the path of the particle from its equatorial position P to the mirror point Q, in units of ka, the distance from the dipole to P; in the same units r_t is the distance from Q to the dipole, and z_t is the arc length of the line of force from P to Q; D is the latitude of the mirror point. (Wentworth, MacDonald and Singer, 1959, p. 502).

point. The value of s has been calculated for some paths by Wentworth, MacDonald and Singer (1959); see Fig. 3 (their Fig. 2). For example, for $\sin \epsilon_0 = 0.1$, or $\epsilon_0 = 5°.7$, s is approximately given by

$$s = 1.25ka \tag{5.51}$$

A particle with this equatorial pitch angle, on the field line $k = 6$, would have a mirror point 3190 km above the ground, in latitude approximately 60°, and the field strength there would be $14{,}800\gamma$. Its oscillation time T_0 is $1.91 \times 10^{10}/v$, if v is expressed in cm/sec. Where this field line crosses the equatorial plane, B is approximately 148γ.

TABLE 5.1

Aspects of the motion of electrons and protons of various energies moving in the field of the earth dipole, and crossing the equatorial plane with the pitch angle 5°.7 at six earth radii from the earth's center; charge $\pm e$, $e = 4.8025 \times 10^{-10}$ esu $= 1.602 \times 10^{-20}$ emu.

(In this Table X^y signifies $X \times 10^y$; relativity change of mass is taken into account.)

v cm/sec	E kev	R_M	C_{St} km	R km	T_L sec	T_0 sec	$\delta\phi$ deg	$\frac{360°}{\delta\phi}$	T_R
				Electrons					
1.76^9	0.879	100	9.00^6	6.76^{-2}	2.42^{-4}	10.95	0.0055	65455	8.4 days
5.28^9	7.85	300	5.20^6	2.03^{-1}	2.46^{-4}	3.67	0.0185	19459	19.5 hours
9.95^9	30.7	600	3.67^6	4.06^{-1}	2.47^{-4}	1.92	0.0356	10121	5.4 hours
1.52^{10}	81.4	1,000	2.85^6	6.76^{-1}	2.81^{-4}	1.27	0.0578	6233	2.2 hours
2.84^{10}	1.07^3	5,000	1.27^6	3.38	7.48^{-4}	0.673	0.2845	1265	14.2 minutes
2.96^{10}	2.53^3	10,000	9.00^5	6.76	1.44^{-3}	0.645	0.5673	635	6.8 minutes
				Protons					
4.79^7	1.2	5,000	1.27^6	3.38	0.433	3.99^2	0.2845	1265	6.6 days
9.58^7	4.79	10,000	9.00^5	7.76	0.433	1.99^2	0.5673	635	35 hours
2.87^8	43.1	30,000	5.20^5	2.03^1	0.433	66.6	1.712	210	3.9 hours
5.75^8	172	60,000	3.67^5	4.06^1	0.433	33.2	3.429	105	58 minutes
9.57^8	479	100,000	2.85^5	6.76^1	0.433	20.0	5.688	63.3	21 minutes
9.13^9	4.79^4	1,000,000	9.00^4	6.44^2	0.433	2.09	57.06	6.3	16 seconds

Table 1 (Akasofu and Chapman 1961) illustrates some aspects of the motion of electrons and protons in the dipole field here considered, for a series of values of their speed and energy (expressed in kilo electron volts—kev; 1 kev $= 1.602 \times 10^{-9}$ erg). The corresponding values of the magnetic rigidity R_M and Störmer length unit C_{St} (see 3.65) are also given. The values of R_M are those of an illustrative table given by Störmer (1955, p. 294; a few corrections here made are required in his Table and in the original of Table 1). The Larmor radius R and period T_L are given for such particles moving wholly in the equatorial plane, namely for $\epsilon_0 = 90°$ and $w = v$, in (13), at 6 earth radii; for other pitch angles w and R are less; elsewhere on the same line of force B is greater, hence R and T_L are reduced.

The table also gives values of T_0, the time of a complete oscillation in latitude, for $k = 6$ and $\sin \epsilon_0 = 0.1 = 5^\circ.7$; the ratios $T_0/T_L(= 5aB_0/2\pi k^2 R_M)$ range from about 4 to about 800 for the protons, and for the electrons from about 400 to about 40,000.

The next column of Table 1 gives the displacement in longitude of a particle that starts from the equatorial plane on the field line $k = 6$ and makes a complete oscillation. It is based on the drift speed, given by the last two terms in (28). These two terms, by (22), (21), and (37), are equal to

$$-(mc\mathbf{d}/eB\rho)(\tfrac{1}{2}w^2 + u^2), \tag{5.52}$$

because for the dipole field curl $\mathbf{B} = 0$ (but if the field of the drift current appreciably modifies the dipole field, curl $\mathbf{B} \neq 0$). The motion is eastward for the electrons, and westward for the protons. As $\frac{1}{2}w^2 + u^2 = v^2(1 - \frac{1}{2}\sin^2\epsilon)$, the drift speed (52) may also be written

$$-(mcv^2/eB\rho)(1 - \tfrac{1}{2}B/B_m), \tag{5.53}$$

using (49). This varies along the path, and at polar angle θ it may be expressed as $r\dot{\phi}\sin\theta$. Denoting an element of the line of force by ds, so that $ds = udt = (v \cos\epsilon)dt$, one may write

$$r\dot{\phi}\sin\theta = \sin\theta\frac{d\phi}{d\theta}\frac{rd\theta}{ds}\frac{ds}{dt} = \frac{v\sin^2\theta\cos\epsilon}{C}\frac{d\phi}{d\theta}, \tag{5.54}$$

because $r\, d\theta/ds = H/F$ (in the notation of 1.8) $= \sin\theta/C$. Hence, considering only the magnitude of $d\phi/d\theta$ (because the sign of the drift, and hence of the change in ϕ, has already been indicated),

$$\frac{d\phi}{d\theta} = \frac{C(r\dot{\phi}\sin\theta)}{v\sin^2\theta\cos\epsilon} = \frac{mcvC(1 - \frac{1}{2}B/B_m)}{eB\rho\sin^2\theta(1 - B/B_m)^{1/2}}, \tag{5.55}$$

using (49). Also, using (3.65), and (1.12),

$$\frac{mcvC}{eB\rho\sin^2\theta} = \frac{k^3a^3\sin^4\theta}{C_{St}^2\rho} = \frac{3k^2a^2\sin^3\theta(1+\cos^2\theta)}{C_{St}^2C^3}$$

Thus the change $\delta\phi$ of ϕ in a complete oscillation in latitude is given by

$$\delta\phi = 4\left(\frac{ka}{C_{St}}\right)^2 I_1(\theta_m),$$

where

$$I_1(\theta_m) = \int_{\theta_m}^{\pi/2}\frac{3\sin^3\theta(1+\cos^2\theta)(1-\frac{1}{2}B/B_m)}{(1+3\cos^2\theta)^{3/2}(1-B/B_m)^{1/2}}\,d\theta$$

Here $B/B_m = \sin^2\epsilon = (C/C_m)(\sin\theta/\sin\theta_m)^6$, by (47, 49).

Alfvén (1940; also 1950, p. 28), to whom this calculation is due, gave the diagram for I_1 here reproduced as Fig. 4. It shows that I_1, there given as a function of the latitude $\phi_0(= 90 - \theta_m)$ of the mirror point, ranges from about 63° for small mirror latitudes, to about 76° for mirror latitude 60°, beyond which it would seem that the value is nearly constant. Hence for mirror latitudes of 60° and above we may write

$$\delta_{\phi} = 4 \times 76°(ka/C_{St})^2, \qquad (5.56)$$

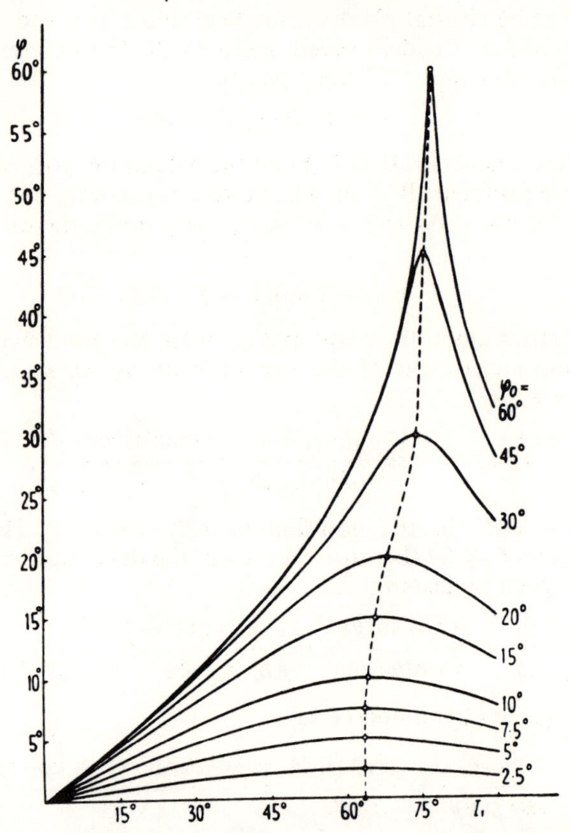

Fig. 5.4. Connexion between displacement in longitude (proportional to I_1) and latitude φ for a particle oscillating through the equatorial plane with amplitude φ_m. (Alfvén, 1950, p. 29.)

and the values of $\delta\phi$ in Table 1 are calculated from this formula, for $k = 6$.

The number of latitude oscillations made by the particle during a complete drift round the dipole axis is $360°/\delta\phi$, and this also is tabulated in Table 1, together with T_R, the time of such a revolution round the axis; clearly

$$T_R = (360°/\delta\phi)T_0 \qquad (5.57)$$

For protons of energy greater than 10 Mev, the revolution is rapid; T_R becomes comparable with T_0, the time of oscillation in latitude. For such energies the guiding center approximation ceases to be valid. The Larmor radius R begins to be an appreciable fraction of an earth radius.

5.12. Model calculations of the DR current and field

In the steady state of a plasma distribution in a constant dipole field there is symmetry about the dipole axis, and also relative to the equatorial plane. If the pitch angle distribution has the form (41), as will here, for simplicity, be supposed, and if the speed distribution function at a point P_0 at radial distance ka in the equatorial plane is denoted by

$$n(k)f(v, k), \quad \text{where} \quad \int_0^\infty f(v, k)dv = 1, \tag{5.58}$$

then at a point P at polar angle θ on the field line k the speed distribution function is $n(k, \theta)f(v, k)$, where

$$\mathrm{n}(k, \theta) = n(k)(B_0/k^3B)^{\beta/2} = n(k)(\sin\theta)^{3\beta}/C^{\beta/2}. \tag{5.59}$$

Hence at P the pressures along and normal to $\mathbf{B}$ are given by

$$p_b(k, \theta) = S(k)/(\beta + 3), \quad p_n(k, \theta) = (\beta + 2)S(k)/2(\beta + 3), \tag{5.60}$$

where

$$S(k) = \sum n(k, \theta)m \int_0^\infty v^2 f(v, k)dv = 2 \sum n(k, \theta)E(k); \tag{5.61}$$

the sum is here to be taken over each kind of particle; $E(k)$ denotes the mean kinetic energy of the particles of a set, at any point on the field line k.

The factor $\mathbf{b} \times \nabla p_n$ in the expression (40) for the electric current intensity $\mathbf{i}$ has the following value, when the plasma distribution is, as here, independent of the azimuth angle ϕ;

$$\frac{1}{h_2}\frac{\partial p_n}{\partial k}\mathbf{d}, \tag{5.62}$$

where h_2 is the second factor in the expression

$$\mathbf{ds} = \mathbf{b}h_1 d\theta + \mathbf{c}h_2 dk + \mathbf{d}h_3 d\phi \tag{5.63}$$

for a line element at P. In the dipole field

$$h_2 = (\sin\theta)^3/C. \tag{5.64}$$

Hence $\mathbf{i}$ at P is given by

$$i(k, \theta) = -f_1(\beta, \theta)S(k) - f_2(\beta, \theta)\partial S(k)/\partial k, \tag{5.65}$$

where

$$f_1(\beta, \theta) = 3\beta ck^2(\sin\theta)^{5+3\beta}/2(\beta + 3)B_0 aC^{\beta/4}, \tag{5.66}$$

$$f_2(\beta,\theta) = ck^3(\beta + 2)(\sin\theta)^{3(\beta+1)}/2(\beta + 3)B_0 aC^{\beta/4}. \tag{5.67}$$

If the speed distribution function is the same everywhere, that is, if it is independent of k, then

$$S(k) = 2N(k)E, \tag{5.68}$$

where E now denotes the mean kinetic energy taken over all types of particle present, and $N(k)$ denotes the total number density of particles at P_0. Thus

$$i(k, \theta) = -2E\{f_1(\beta, \theta)N(k) + f_2(\beta, \theta)\partial N(k)/\partial k\} \tag{5.69}$$

Akasofu and Chapman (1961) considered the model Gaussian distribution of plasma given by

$$N(k) = N_0 \exp\{-g^2(k - k_0)^2\}, \tag{5.70}$$

with particular reference to the numerical values:

$$k_0 = 6, \quad \beta = -\tfrac{1}{2}, \quad g = 1.517. \tag{5.71}$$

This model "V3" radiation belt has its maximum density N_0 at radius $6a$; the density is less by the factor 10 at one earth radius on either side of this center line of the belt. The choice of β implies an excess of smaller pitch angles.

They also considered the following values, more nearly, as to size, agreeing with the outer Van Allen ("V2") belt:

$$k_0 = 3.5, \quad \beta = 1, \quad g = 1.517. \tag{5.72}$$

Some of their numerical results for these two belts are as follows (see their equations 90, 91, 73, 75, 83′, 84′, 83, 84): for V2 only the first factor is given, the others being the same as for V3: E is supposed expressed in kev.

Total number of particles	:V3:	$7.31 \times 10^{28} N_0$	;V2:	1.75
Total current	:V3:	$9.38 \times 10^{3} N_0 E$ amperes	;V2:	1.18
Magnetic moment	:V3:	$1.17 \times 10^{23} N_0 E$ gauss cm^3	;V2:	0.0688
DR field at earth's center	:V3:	$-0.306 N_0 E \gamma$	;V2:	−0.0572

The values are less for V2 than for V3 because of the smaller circumference and volume of the V2 belt.

Akasofu and Chapman (1961) also gave the value of the field of the V3 belt along an equatorial radius. They compared it (Fig. 5) with observations of a total-force magnetometer carried by the satellite Explorer VI (Smith, Coleman, Judge and Sonett 1960); the instrument measured not the actual total intensity, but the component normal to the spin axis of the satellite. Thus the observed and calculated curves are not strictly comparable. The observations, of course, referred to a component of the combined field, of the dipole and the radiation belts; the authors cited showed in their curve, reproduced in the upper part of Fig. 5, both the total intensity of the dipole field (the curve marked H theoretical) and their estimate of the component intensity H_n of the actual field along the direction of measurement. The observations are shown by the small circles marked Observed H_n. They suggest a marked reduction of the combined dipole and belt field at $6.5a$, close to the distance at which, according to the calculations of Akasofu and Chapman (1961), their model V3 belt would produce a minimum in the combined field. This minimum is slightly further out than the minimum of the DR field, shown in the lowest section of Fig. 5. Their choice of $6a$ as the mean radius of their V3 belt was made to fit the field line that enters the atmosphere in the auroral zone. In Fig. 5, lowest section, a numerical value (150) is chosen for N_0E, to make their combined

H curve resemble the observed H_n curve. Also, in the illustrations of the density and current intensity along an equatorial radius, a value was chosen for N_0; the latter, taken to be $1/\text{cm}^3$, then corresponds to $E = 150$ kev.

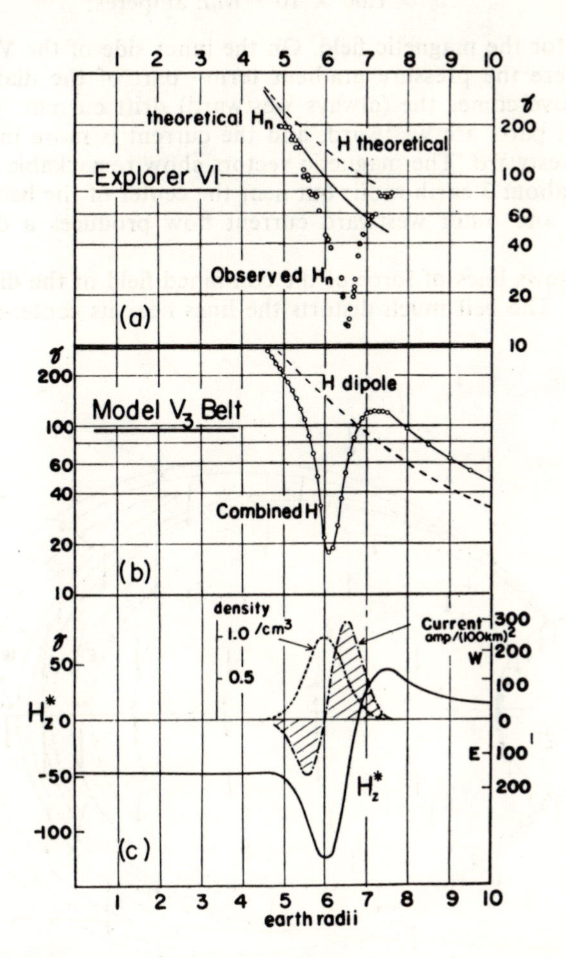

Fig. 5.5. (*a*) The magnetic field measurements H_n made by the magnetometer carried on Explorer VI (Smith, Coleman, Judge and Sonett, 1960). (*b*) The field distortion produced in the equatorial plane by the model V_3 belt indicated in (*c*). (*c*) The number density distribution $N(k)$ for the model belt V_3, and the corresponding distribution of current intensity $i(k)$ and magnetic intensity $B(k)$, for the equatorial plane. (Akasofu and Chapman, 1961, p. 1336.)

Akasofu, Cain and Chapman (1961, 1962) have published the results of much more extensive calculations based on the above (and derivative) formulae given by Akasofu and Chapman (1961). They give the current distribution and its DR field not only along the equatorial radius but over the meridian plane. Figure 6 shows such results for the V3 belt, by isolines for the current intensity, or rather for i/i_0, where

$$i_0 = 1.66 \times 10^{-15} N_0 E \text{ amperes:}$$

and by vectors for the magnetic field. On the inner side of the V3 belt the current is eastward; there the pressure gradient term—part of the diamagnetic field—is eastward, and overcomes the (always westward) drift current; on the outer side of the belt both parts are westward, and the current is more intense, so that the net current is westward. The magnetic vectors show remarkable uniformity of the *DR* field up to about 3 earth radii; but near the center of the belt cross section the inner eastward and outer westward current flow produces a deep minimum of magnetic force.

Figure 7 shows lines of force of the combined field of the dipole and the belt, for $N_0E = 150$. The belt much distorts the lines near its center-line.

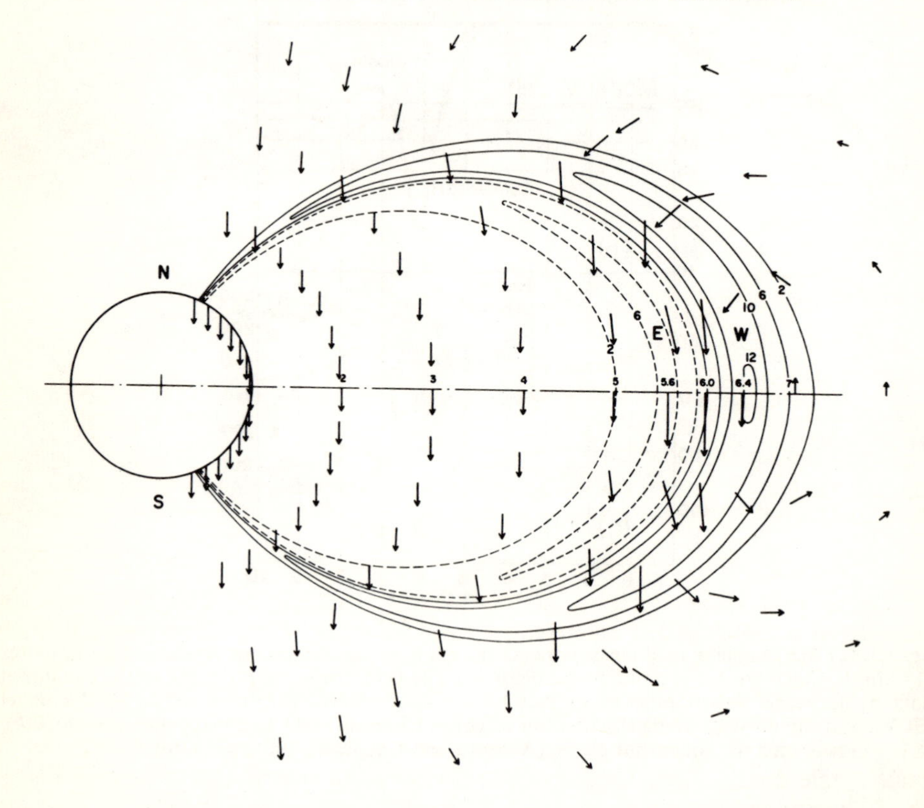

Fig. 5.6. The DR field vectors, for the first approximation to the magnetic field of the model belt in a meridian plane; and isolines of the equivalent current intensity in the belt (solid lines indicate westward, dashed lines eastward current). The vector scale of force, and the unit in which the current intensity is expressed, are proportional to the energy density N_0E (kev/cm³) at the center line of the belt, at 6 earth radii from the earth's center.(Akasofu, Cain and Chapman, 1961, p. 4017.)

These calculations of the belt field are usefully illustrative, but they are only approximate, because the particle motions are calculated without taking account of the field they themselves collectively produce. Akasofu, Cain and Chapman (1961) made a second approximation to the belt field, by calculating i from the motions of the particles in the combined field of the dipole, and of the first approximation to the belt field, shown in Fig. 6. The resulting change in the calculated field within about 3 earth radii was less than about 5%, but within the cross section of the belt the minimum was deepened, by nearly 20%. There is need to develop a selfconsistent treatment of the problem, but so far the difficulties of this task have been exposed (Beard, 1962; Akasofu 1962) rather than overcome.

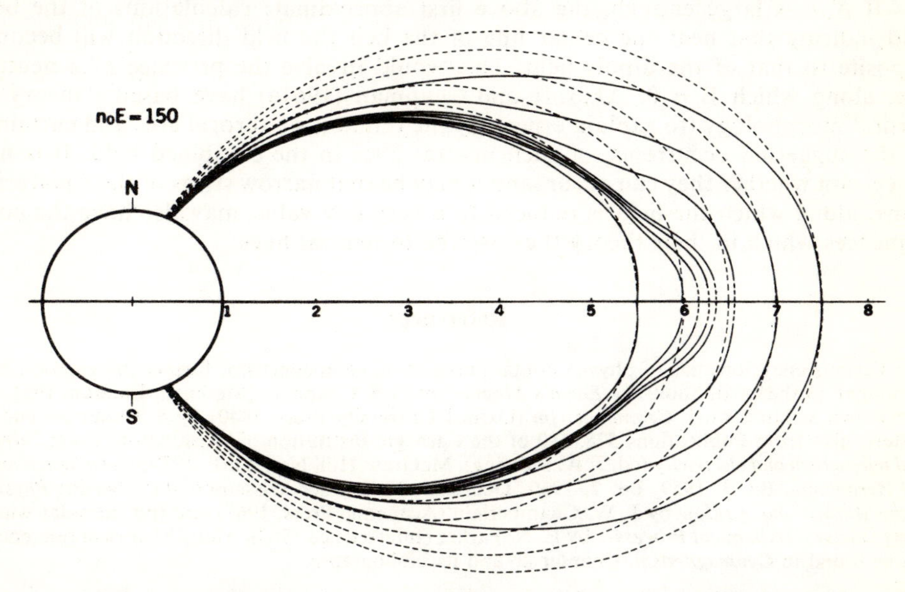

Fig. 5.7. The lines of force of the first approximation to the combined field of the geomagnetic dipole and the belt. The dashed lines are lines of force of the dipole field. (Akasofu, Cain and Chapman, 1961, p. 4022.)

Akasofu, Cain and Chapman (1962; see also Akasofu and Cain 1962) have also considered a rather more complicated form of model belt, corresponding to the choice of different values of g in (66), for values of k less and greater than k_0. This enables an approximate fit to be made with belts such as the one composed of relatively low energy protons (150 kev to 4.5 Mev) revealed by the satellite Explorer XII (Davis and Williamson 1962); these observations also gave an indication of the pitch angle distribution. Consequently Akasofu, Cain and Chapman (1962) took the following values:

$$k_0 = 3.2, \quad \beta = 2.0, \cdot g = 2.99(k < k_0), \quad g = 0.419(k > k_0). \tag{5.73}$$

$$N_0 = 0.2/\text{cm}^3, \quad E = 500 \text{ kev}, \quad N_0E = 100. \tag{5.74}$$

This belt well fits the density distribution of the one observed by Davis; the gradient is much steeper on the inner than on the outer side; its limits (if taken to be where $N = 0.001/cm^3$) are at $2.4a$ and $8.7a$. The adopted value of β gives a rather good fit to the observed pitch angle distribution. The maximum calculated belt field intensity is -24γ, at $4.1a$; the minimum is rather broad. Thus at $4.1a$ the dipole field is reduced by 5%. Beyond $7.2a$ the belt increases the dipole field. At the earth the field is -13γ.

Such calculations are among the primitive beginnings of the theoretical exploration of the electric current and *DR* field distribution in the region of the Van Allen belts. During magnetic storms rapid and complicated changes must occur in this region.

If N_0E is large enough, the above first approximate calculations of the belt field indicate that near the center line of the belt the field direction will become opposite to that of the dipole field. This would involve the presence of a neutral line, along which $\mathbf{B} = 0$. Akasofu and Chapman (1961a) have based a theory of auroral morphology (to explain especially the thinness of auroral arcs and curtains) on the suggested occurrence of such neutral lines in the combined field. It is not yet certain whether they can occur, and it may be that narrow strips of the equatorial plane, along which the field is reduced to a very low value, may also have the consequences which in their theory they ascribe to neutral lines.

References

Various encyclopedias of physics contain articles on geomagnetism; further information can be sought in the small book *The Earth's Magnetism* by S. Chapman (Methuen, London, 1951) or in the two volume work *Geomagnetism* (Oxford University Press, 1940) by S. Chapman and J. Bartels; also from Publications 578, 580 of the Carnegie Institution of Washington, 1947: *Terrestrial magnetism and electricity* (ed. J. A. Fleming), McGraw-Hill, New York, 1939; *Geomagnetismus und Aeronomie*, Bd. 3, 1959, Bd. 2, 1960, Deutscher Verlag der Wissenschaften, Berlin; *Physics of the Aurora and Airglow* by J. W. Chamberlain (Academic Press, 1961) and (on the solar wind) *Interplanetary Dynamical Processes* by E. N. Parker (Interscience 1963). Many historical references can be found in *Geomagnetism*, Chapter 26 and its bibliography.

Adams, W. G., 1892, Comparison of simultaneous magnetic disturbances at several observatories, *Phil. Trans. Roy. Soc.* (A), **183**, 131–140.

Akasofu, S. -I., 1962, On a self-consistent calculation of the ring current field, *J. Geophys. Res.*, **67**, 3617–3618.

Akasofu, S. -I. and J. C. Cain, 1962, The magnetic field of the radiation belts, *J. Geophys. Res.*, **67**, 4078–4080.

Akasofu, S. -I., J. C. Cain and S. Chapman, 1961, The magnetic field of a model radiation belt, numerically computed, *J. Geophys. Res.*, **66**, 4013–4026.

Akasofu. S. -I., Cain, J. C., and S. Chapman, 1962, The magnetic field of the quiet-time proton belt, *J. Geophys. Res.*, **67**, 2645–2647.

Akasofu, S. -I. and Chapman, S., 1960a, Some features of the magnetic storms of July 1959, and tentative interpretation, *UGGI Monograph No.* 7, 93–108.

Akasofu, S. -I. and Chapman, S., 1961, The ring current, geomagnetic disturbance, and the Van Allen radiation belts, *J. Geophys. Res.*, **66**, 1321–1350.

Akasofu, S. -I. and Chapman, S., 1961a, A neutral line discharge theory of the aurora polaris, *Phil. Trans. Roy. Soc.*, London, (A) **253**, 359–406.

Akasofu, S. -I. and Chapman, S., 1962, Large-scale auroral motions and polar magnetic disturbances—III. The aurora and magnetic storm of 11 February 1958. *J. Atmos. and Terr. Phys.*, **24,** 785–796.

Alfvén, H., 1950, *Cosmical Electrodynamics*. (Oxford University Press).

—— 1956, On the theory of sunspots, *Tellus*, **8**, 274–275.

Alfvén, H., 1958, On the theory of magnetic storms and aurorae, *Tellus*, **10**, 104–116.
—— 1960, On the motion of a charged particle in a magnetic field, *Arkiv. Mat. Astr. Fys.*, **27A**, No. 22.
Aller, L. H., 1954, Spectrographic studies of the combination variables Z Andromedae, BF Cygni and CI Cygni, *Pub. Dom. Ap. Obs.*, **9**, 321–362.
Allen, C. W., 1960, A sunspot cycle model, *Observatory*, **80**, 94–98.
Apel, J. R. and Singer, S. F., 1961, Magnetic effects of trapped particles (Abstract), *J. Geophysical Res.*, **66**, 2510.
Axford, W. I. and C. O. Hines, 1961, A unifying theory of high-latitude geophysical phenomena and geomagnetic storms, *Canadian J. Phys.*, **39**, 1433–1464.
Axford, W. I., J. A. Fejer and C. O. Hines, 1962, A comprehensive model of auroral and geomagnetic disturbances, *J. Phys. Soc. Japan, Supplement A-I*, 173–175.
Axford, W. I. and G. C. Reid, 1962, Polar-cap absorption and the magnetic storm of February 11, 1958, *J. Geophys. Research*, **67**, 1692–1696.
Babcock, H. W., 1953, The solar magnetograph, *Astrophys. J.*, **118**, 387–396.
—— 1961, The topology of the sun's magnetic field and the 22-year cycle, *Astrophys. J.*, **133**, 572–587.
Baker, W. G. and D. F. Martyn, 1953, Electric currents in the ionosphere, I. The conductivity, *Phil. Trans. Roy. Soc.*, London, (A) **246**, 281–294.
Bartels, J., 1932, Terrestrial magnetic activity and its relations to solar phenomena, *Terr. Magnetism and Atmos. Elec.*, **37**, 1–52.
—— 1934, Twenty-seven day recurrences in terrestrial-magnetic and solar activity, 1923–33, *Terr. Magnetism and Atmos. Elec.*, **39**, 201–202.
—— 1957, The geomagnetic measures for the time-variations of solar corpuscular radiation, described for use in correlation studies in other geophysical fields, *I.G.Y. Annals*, **4**, 227–236.
Beard, D. B., 1960, The interaction of the terrestrial magnetic field with the solar corpuscular radiation, *J. Geophys. Res.*, **65**, 3559–3568.
—— 1962a, *Ibid.*, Part 2, *J. Geophys. Res.*, **67**, 477–483.
—— 1962b, Self-consistent calculation of the ring current, *J. Geophys. Res.*, **67**, 3615–3616.
Berkowitz, J. and C. S. Gardner, 1957, On the asymptotic series expansion of the motion of a charged particle in slowly varying fields. (New York University Report NYO-7975).
Biermann, L., 1951, Kometenschweife und solare Korpuskularstrahlung, *Zs. f. Astrophys.*, **29**, 274–286.
—— 1961, The solar wind and the interplanetary media, In *Space Astrophysics*, ed. W. S. Liller. (New York: McGraw-Hill).
Birkeland, Kr., 1908, The Norwegian Aurora Polaris Expedition, 1902–1903, vol. 1, *Christiania;* (H. Aschehoug & Company).
Bogolyubov, N. N. and D. N. Zubarev, 1955, An asymptotic approximation method for a system with rotating phases and its application to the motion of a charged particle in a magnetic field, *Ukrainian Math. J.*, VII, 5.
Bruce, C. E. R., 1956, Les étoiles variables à longue période à spectre composite, *Comptes rendus* (Paris), **242**, 2101–2103.
—— 1957a, Gas movements in the long-period variables. *Observatory*, **77**, 107–108; 153–155.
—— 1957b, Excitation and gas movements in the long period variable stars, *Mèm. Soc. R. Liége*, **20**, 364–369.
—— 1959, Cosmic thunderstorms, *J. Franklin Inst.*, **268**, 425–445.
—— 1960a, Un thermomètre cosmique à vitesse de gaz. *Comptes rendus* (Paris), **250**, 61–63.
—— 1960b, Temperatures reached in electrical discharges in the solar atmosphere, *Nature*, **187**, 865.
—— 1961, Spiral stellar nebulae and cosmic gas jets. *J. Franklin Inst.*, **271**, 1–11.
—— 1962a, P Cygni, γ Cassiopeiae and the planetary nebulae, *J. Franklin Inst.*, **274**, 284–299.
—— 1962b, Stellar temperatures, *Observatory* (in press).
—— 1962a, The gas pressure in cosmic electrical discharges. *J. Franklin Inst.*, **273**, 59.
Chamberlain, J. W., 1961, Interplanetary gas, III, A hydrodynamic model of the corona, *Astrophys. J.*, **133**, 675–687 (and earlier papers there cited).
Chapman, S., 1918, An outline of a theory of magnetic storms, *Proc. Roy. Soc.*, (A) **95**, 61–83.
—— 1929, Solar streams of corpuscles; their geometry, absorption of light and penetration, *Monthly Notices Roy. Astron. Soc.*, **89**, 456–470.

Chapman, S., 1935, The electric current systems of magnetic storms, *Terr. Magnetism & Atmos. Elec.*, **40**, 349–370.

—— 1956, The electrical conductivity of the ionosphere: a review, *Nuovo Cimento Supplemento al* Vol. 4, Serie X, 1385–1412.

—— 1956, The morphology of geomagnetic storms and bays, a general review, *Vistas in Astronomy*, **2,** 912–928. Pergamon Press, London.

—— 1960, Idealized problems of plasma dynamics relating to geomagnetic storms, *Rev. Mod. Phys.*, **32,** 919–933.

Chapman, S. and J. Bartels, 1940, *Geomagnetism*, 2 vols. (Oxford: Clarendon Press).

Chapman, S. and Ferraro, V. C. A., 1930, A new theory of magnetic storms, *Nature*, **126**, 129–130.

—— 1931, 1932, 1933, A new theory of magnetic storms, *Terr. Magnetism & Atmos. Elec.*, **36**, 77–97, 171–186; **37**, 147–156, 421–429; **38**, 79–96.

—— 1940, The theory of the first phase of a geomagnetic storm, *Terr. Magnetism & Atmos. Elec.*, **45**, 245–268.

Chapman, S. and P. C. Kendall, 1961, An idealized problem of plasma dynamics that bears on geomagnetic storm theory; oblique projection, *J. Atmos. and Terr. Phys*, **22**, 142–156.

Chree, C. 1908, Magnetic declination at Kew Observatory 1890 to 1900, *Phil. Trans. Roy. Soc.* (A), **208**, 205–246.

—— 1912, 1913, Some phenomena of sunspots and terrestrial magnetism at Kew Observatory, *Phil. Trans. Roy. Soc.* (A), **212**. 75–116; **213**, 245–277.

Chree, C. and Stagg, J. M., 1927, Recurrence phenomena in terrestrial magnetism, *Phil. Trans. Roy. Soc.* (A) **227**, 21–62.

Christiansen, W. N. and others, 1960, A study of a solar active region using combined optical and radio techniques, *Ann. d'Ap.*, **23**, 75–101.

Cole, K. D., 1960, A dynamo theory of the aurora and magnetic disturbance, *Australian J. Phys.* **13,** 484–497.

Davis, L. R. and J. M. Williamson, 1962, Low energy trapped protons; Paper presented at the 3rd International Space Science Symposium (to be published).

Dellinger, J. H., 1937, Sudden disturbances of the ionosphere, *Proc. I.R.E. Inst. Radio Eng.* **25**, 1253–1290.

Dessler, A. J. and E. N. Parker, 1959, Hydromagnetic theory of geomagnetic storms, *J. Geophys. Res.*, **64**, 2239–2252.

Dungey, J. W., 1958, *Cosmic Electrodynamics*. (Cambridge University Press).

—— 1961, The steady state of the Chapman-Ferraro problem in two dimensions, *J. Geophys. Res.*, **66**, 1043–1047,

Ellison, M. A., 1956, *The Sun and its Influence*. (London: Routledge & Paul).

Ferraro, V. C. A., 1952, On the theory of the first phase of a geomagnetic storm; a new illustrative calculation based on an idealized (plane not cylindrical) field distribution, *J. Geophys. Res.*, **57**, 15–49.

—— 1960, An approximate method of estimating the size and shape of the stationary hollow carved out of a neutral ionized stream of corpuscles impinging on the geomagnetic field, *J. Geophys. Res.*, **65**, 3951–3953.

Finch, H. F. and Leaton, B. R., 1957, The earth's main magnetic field-Epoch 1955.0, *Monthly Notices Roy. Astron. Soc., Geophysical Supplement*, **7**, 314–317.

Fried, B. D., 1960, Translation of Bogolyubov and Zubarev 1955. (Space Technology Laboratories, Inc.).

Friedman, H., 1960, The sun's ionizing radiations, In *Physics of the Upper Atmosphere*, ed. J. A. Ratcliffe. (New York: Academic Press, pp. 133–218).

Gold, T., 1959, Motions in the magnetosphere of the earth, *J. Geophys. Research*, **64,** 1219–1224,

Gold, T., 1962, Magnetic storms, *Space Science Reviews*, **1,** 100–114.

Greaves, W. M. H. and Newton, H. W., 1929, On the recurrence of magnetic storms, *Monthly Notices Roy. Astron. Soc.*, **89**, 641–646.

Greenwich, 1955, *Sunspots and Geomagnetic Data*, 1874–1954. (London: H. M. Stationery Office).

Hale, G. E., 1931, The spectrohelioscope and its work, III, Solar eruptions and their apparent terrestrial effects, *Astrophys. J.*, **73**, 379–412.

Hulburt, E. O., 1929, On the ultraviolet light theory of aurorae and magnetic storms, *Phys. Rev.*, **34,** 344–351, 1167–1183; 1930, *ib.*, **36,** 1560–1569.

Hulburt, E. O., 1937, Terrestrial magnetic variations and aurorae, *Rev. Mod. Phys.*, **9,** 44–68.

Hulburt, E. O., 1954, Magnetic storms, aurorae, ionosphere and zodiacal light, *Sci. Monthly*, **78,** 100–109.

Hurley, J., 1961a, Interaction of a streaming plasma with the magnetic field of a line current. *Phys. Fluids*, **4**, 109–111.

—— 1961b, Interaction of a streaming plasma with the magnetic field of a two-dimensional dipole. *Phys. Fluids*, **4**, 854–859.

Kahn, F. D., 1949, An investigation into the possibility of observing streams of corpuscles emitted by solar flares. *Monthly Notices Roy. Astron. Soc.*, **109**, 324–336.

Kantrovitz, A. R., 1960, Contribution to discussion, Magneto-fluid dynamics, *Rev. Mod. Phys.*, **32**, 939.

Kruskal, M., 1960, In *La Théorie des Gaz Neutres et Ionisés* eds. C. DeWitt and J. -F. Detoeuf, (Paris: Hermann; New York; John Wiley).

Larmor, J., 1895, A dynamical theory of the electric and luminiferous medium: Part II, Theory of electrons, *Phil. Trans. Roy. Soc.*, London, (A), **186**, 695–743; see also *Aether and Matter* (Cambridge University Press, 1900), p. 227.

—— 1897, On the theory of magnetic influences on spectra; and on the radiation from moving ions, *Phil. Mag.*, **44**, 503–512. Collected Works **2**, 149. (Cambridge University Press, 1929).

Lebeau, A. and Schlich, R., 1962, Sur une propriété de l'activité magnétique diurne dans les régions de haute latitude (Stations Charcot et Dumont d'Urville). Comptes Rendus, Paris, **254**, 3014–3016.

Lindemann, F. A., 1919, Note on the theory of magnetic storms, *Phil. Mag.* **38**, 669–684.

Lüst, R. and Schlüter, A., 1955, Drehimpulstransport durch Magnetfelder und die Abbremsung rotierender Sterne, *Zs. f. Astrophys*, **38**, 190–211.

Martyn, D. F., 1951, The theory of magnetic storms and auroras, *Nature*, **167**, 92–94.

Maunder, E. W., 1904, 1905, 1916, Magnetic disturbances . . . at Greenwich, and their association with sunspots, *Monthly Notices Roy. Astron. Soc.*, **64**, 205–224; **65**, 2–34, 538–559, 666–681; **76**, 63–68.

—— 1922, The sun and sunspots, 1820–1920, *Monthly Notices Roy. Astron. Soc.*, **82**, 534–542.

Maurain, C., 1934, Sur l'intervalle de temps entre les phénomènes solaires et les perturbations magnétiques terrestres, *Annales de l'Inst. Phys. du Globe, Paris*, **196**, 1182–1186.

Maxwell, J. C., 1873, *A Treatise on Electricity and Magnetism*. (Cambridge University Press). 2nd edition 1881, 3rd edition 1891, Dover edition 1954; see p. 657, 2nd edition.

Mayaud, P. N., 1949, Activité magnétique dans les régions polaires. Expéditions polaires françaises, *Résultats Scientifiques*, S. IV 2.

Midgley, J. E. and L. Davis, Jr., 1962, Computation of the bounding surface of a dipole field in a plasma by a moment technique, *J. Geophys. Res.*, **67**, 499–504.

Milne, E. A., 1926, On the possibility of the emission of high-speed atoms from the sun and stars, *Monthly Notices Roy. Astron. Soc.*, **86**, 459–473.

Newton, H. W., 1958, *The Face of the Sun*. (Penguin Books, 208 pp.).

Northrop, T. G., 1961, The guiding center approximation to charged particle motion, *Ann. Phys.*, **15**, 79–101.

Obayashi, T. and Y. Hakura, 1960, Enhanced ionization in the polar ionosphere and solar corpuscular radiation, In *Space Research*. (Amsterdam: North-Holland Publishing Co., pp. 665–694).

Parker, E. N., 1957, Newtonian development of the hydromagnetic properties of ionized gases of low density, *Phys. Rev.*, **107**, 924–933.

—— 1961, The solar wind, In *Space Astrophysics*, ed. W. S. Liller. (New York: McGraw-Hill, pp. 157–170).

Piddington, J. H., 1960, Geomagnetic storm theory, *J. Geophys. Res.*, **65**, 93–106.

Priester, W. and Cattani, D, 1962, On the semi-annual variation of geomagnetic activity and its relation to the solar corpuscular radiation, *J. Atmos. Sci.*, **19**, 121–126.

Ratcliffe, J. A., 1960, *Physics of the Upper Atmosphere*. (New York: Academic Press, pp. xi, 586).

Sabben, D. Van, 1961, Ionospheric current systems of ten IGY-solar flare effects, *J. Atmos. and Terr. Phys.*, **22**, 32–42.

Schmidt, A., 1924/5, Das erdmagnetische Aussenfeld, *Zs. f. Geophysik*, **1**, 1–13.

Silsbee, H. C. and E. H. Vestine, 1942, Geomagnetic bays, their frequency and currents system, *Terr. Mag.*, **47**, 195–208.

Singer, S. F., 1957, A new model of magnetic storms and aurorae, *Trans. Am. Geophys. Union*, **38**, 175–190.

Slutz, R. J., 1962a, The shape of the geomagnetic field boundary under uniform external pressure, *J. Geophys. Res.*, **67**, 505–513.

—— 1962b, Solar wind distortion of the geomagnetic field boundary, *National Academy of Sciences*, U.R.S.I. meeting, April 30, 1962. (To appear in *J. Geophys. Res.*).

Smith, E. J., P. J. Coleman, D. L. Judge and C. P. Sonett, 1960, Characteristics of the extra-terrestrial current system: Explorer VI and Pioneer V, *J. Geophys. Res.*, **65**, 1858–1861.

Sonett, C. P.,E. J. Smith, D. L. Judge, and P. J. Coleman, Jr., 1960, Current systems in the vestigial geomagnetic field: Explorer VI, *Phys. Rev. Letters*, **4**, 161–163.

Spreiter, J. R. and B. R. Briggs, 1962a, Theoretical determination of the form of the boundary of the solar corpuscular stream produced by interaction with the magnetic dipole field of the earth, *J. Geophys. Res.*, **67**, 37–51.

—— 1962b, On the choice of the condition to apply at the boundary of the geomagnetic field in the steady-state Chapman-Ferraro problem, *J. Geophys. Res.*, **67**, 2983–2985.

Stagg, J. M., 1928, The time interval between magnetic disturbance and associated sunspot changes, *Geophysical Memoirs*, **5**, No. 42, 16. (London, Meteorological Office).

—— 1937, Non-instrumental auroral observations, British Polar Year Expedition, Fort Rae, N. W. Canada, 1932–33, *Roy. Soc., London*, **1**, 251–298.

Steenbeck, M., 1961, Skizze zu einer magnetohydrodynamischen Theorie der Sonnenflecken-Vorgänge, *Beiträge aus der Plasmaphysik*, **3**, 153–178.

Störmer, C., 1955, *The Polar Aurora*. (Oxford University Press).

Stratton, J. A., 1941, *Electromagnetic Theory*. (New York: McGraw-Hill).

Sugiura, M. and S. Chapman, 1960, The average morphology of geomagnetic storms with sudden commencement, *Abh. Akad. Wiss. Göttingen, Math.-Phys. K*1. *Sonderheft Nr*. 4, 53 pp.

Tandon, J. N., 1956, Geomagnetic activity and solar *M*-regions for the current epoch of the sunspot minimum, *Indian J. Phys.*, **30**, 153–168.

Van Allen, J. A. and Frank, F. A., 1959, Survey of radiation around the earth to a radial distance of 107,400 kilometers, *Nature*, **183**, 430–434.

Vernov, S. N., Chudakov, A. E., Vakulov, P. V. and Logachev, Yu I., 1959, Study of terrestrial corpuscular radiation and cosmic rays during flight of the cosmic rocket, *Doklady, Akad. Nauk S.S.S.R.*, **125**, 304–307.

Vestine, E. H. and W. L. Sibley, 1940, The disturbance field of magnetic storms, Bull. No. 11, Trans. Washington Meeting, Assoc. of Terr. Magn. and Elec., Int. Union of Geodesy and Geophys., 360–381.

—— 1960, *Geomagnetic field lines in space*. Rand Corporation Report R–368.

Waldmeier, M., 1961, *The sunspot-activity in the years* 1610–1960. (Zurich: Schulthess).

Walén, C., 1949, *On the vibratory rotation of the sun*. (Stockholm: Lindsåthl).

Wentworth, R. C., W. M. MacDonald, and S. F. Singer, 1959, Lifetimes of trapped radiation belt particles determined by Coulomb scattering, *Phys. Fluids*, **2**, 499–509.

Zhigulev, V. N., 1958, Theory of the magnetic boundary layer, English translation: *Soviet Physics, Doklady*, **4**, 57–60, 1959.

—— 1959, On the phenomenon of magnetic "detachment" of the flow of a conducting medium, English translation: *Soviet Physics, Doklady*, **4**, 514–516, 1959.

Zhigulev, V. N. and E. A. Romishevskii, 1959, Concerning the interaction of currents flowing in a conducting medium with the earth's magnetic field, English translation: *Soviet Physics, Doklady*, **4**, 859–862 1959.

Name and place index

Place names are in italics

Dates, historical, or series of data

Dates: geophysical events

Subject Index